ENGINEERING INSIGHTS: CLOUD-ENABLED BI FOR CRITICAL INDUSTRY TRANSFORMATION

Sethu Sesha Synam Neeli

Made with ❤ on the Notion Press Platform
www.notionpress.com

PREFACE

The advancement in technology has revolutionized sectors by reposing new paradigms of operation, competition and growth. Of these emerging technologies, Cloud BI has emerged as the disruptive innovation that addresses the conversion of data into actionable insights. Engineering Insights: Cloud-Enabled BI for Critical Industry Transformation is my effort to explain this powerful suite and its potential across the industries.

This book is a valuable read for technology managers, data lovers, and decisionmakers trying to make sense of the digital world. It gives a general perspective of Cloud BI and features its basics, benefits, and case studies within the healthcare, retail, manufacturing and the financial sectors. In this blog post, I want to showcase how Cloud BI can solve problems, open new doors and create value through examples from different industries.

It has been quite a journey writing this book and at the same time full of hardships and triumphs. I have therefore aimed at providing practical knowledge and different strategies which can help the readers to adopt the Cloud BI not only for adoption but also for its sustainable use.

DEDICATION

My greatest strength has been my family's constant support and encouragement, to whom this book is dedicated. To my parents and grandparents, whose sacrifices, love, and knowledge have paved the way for everything I have accomplished. To my mentor, whose guidance and inspiration have been a beacon of light throughout my journey. Your belief in me has been my driving force, and this work stands as a testament to your profound impact on my life. Thank you for being my source of strength, inspiration, and endless motivation.

ACKNOWLEDGEMENT

It practically impossible to present this book without support and contribution from various individuals and organizations.

To begin with, I express my deep appreciation to my family and friends for their support which inspired me to work on this project. They believed in what I wanted to do and I got the courage to push through.

To my professional peers and teachers, thanks for contributing your ideas and wisdom that influenced the width and correctness of this book. They appreciated the fact that your experiences and perspectives enriched the content of this course immensely.

Particular thanks should go to the leaders and first movers in the Cloud BI market whose efforts informed this project. It is fulfilling and enriching to explore this technology area with your key contributions.

Last but not least, for the readers of this book, thank you for always willing to explore and gaining your knowledge. I also expect that this book will be helpful as you work towards learning about and applying Cloud BI solutions.

With sincere appreciation,

[Sethu Sesha Synam Neeli]

CONTENTS

INTRODUCTION TO CLOUD BUSINESS INTELLIGENCE (BI)

1.1 Chapter Overview

This chapter offers the first overview of Cloud Business Intelligence (BI) and outlines the importance of the subject in the contemporary world of business analytics. BI is defined first and the history of it is reviewed then, leading up to the advent of cloud computing as the next logical step in data analysis tools. The chapter also describes what exactly Cloud BI is and how it uses cloud solutions to provide better data availability, flexibility, and affordability than traditional BI solutions. Main components of the Cloud BI solutions are described to give the audience understanding of their characteristics and purpose. Furthermore, the chapter looks at the various industries, where Cloud BI is being applied to show its versatility. Last but not least, a brief recap of the topics discussed occurs in order to create expectation for further analysis in the subsequent chapters.

1.2 Business Intelligence

Business intelligence combines data visualisation, data mining, best practices, data tools and infrastructure, and business analytics to help businesses make decisions based on data. Having up-to-date business intelligence means you have a full view of your company's data and can use it to make adjustments, eliminate inefficiencies, and respond quickly to changes in supply or the market. Modern business intelligence systems prioritise controllable data on dependable platforms, empowered business users, fast insight, and flexible self-service analysis (Niu et al., 2021).

To solve specific problems faced by businesses, a set of tools called business intelligence (BI) is used. Typically, business intelligence (BI) products are designed to offer a mix of powerful data visualisation tools, analytics platform features, and operational embedded analytics (A-khateeb, 2024).

Tools for statistical analysis, reporting, database management, and data mining make up this category. Many businesses use business intelligence (BI) software independently, either in-house or through a consulting firm or third-party vendor.

The business intelligence platform is mainly a cloud-based solution that helps organisations achieve and maintain the goals of their digital transformation programs. Using a business intelligence platform to store, organise, and analyse data can help businesses gain the insights they need to fulfil their goals.

It may also be used as a tool to help stakeholders and workers alike learn more about how well a firm is performing. There is more to a corporate BI platform than just an analytics tool.

To put it another way, you may utilize a business platform to facilitate communication between all members of the organization and external stakeholders like suppliers, partners, and consumers.

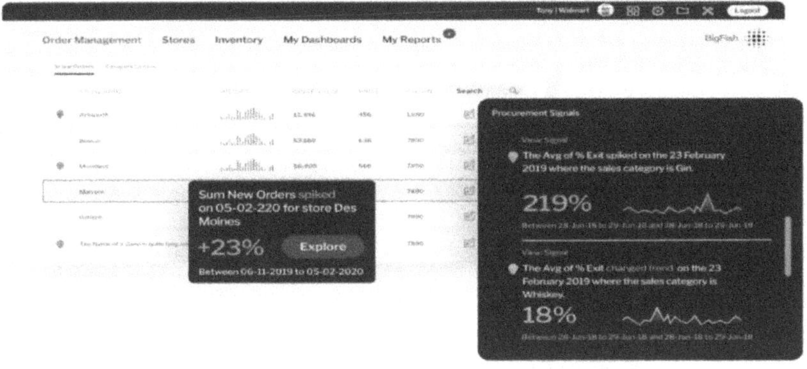

Figure 1.1: the above figure shows a BI dashboard

Source: - *(Team, 2022)*

How business intelligence works

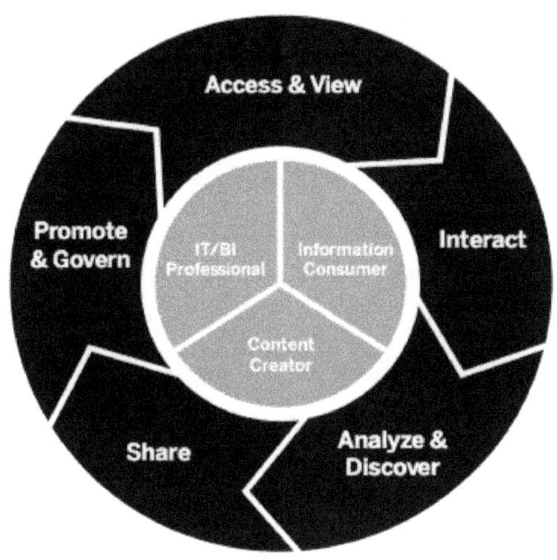

Figure 1.2: The Modern Analytics Workflow

Source: - *(Tableau, 2020)*

The above Figure 1.2 illustrates the "Modern Analytics Workflow," a circular model depicting the different stages involved in the process. Starting from the top, it shows "Access & View," followed by "Promote & Govern," "Share," "Analyze & Discover," "Interact," and finally, "Content Creator." This cyclical nature emphasizes the continuous flow of information and analysis within the workflow.

Businesses and other groups have goals and enquiries. So that they can answer these questions and track their progress towards these goals, they gather the necessary data, analyze it, and then determine the best way to achieve their objectives.

Raw data originates from business processes, technically speaking. Various storage methods, including files, apps, cloud, and data warehouses, are used to store processed data. Once users store the data, they may access it and begin the analytical process to respond business questions.

One feature of business intelligence platforms is data visualization, which lets users create charts and graphs to show data to key stakeholders and decision-makers.

BI methods

Gathering, collecting, and analyzing data from firm operations or activities to maximize performance is the goal of business intelligence, a broad term that includes several procedures and methodologies. Extraordinarily, it's not just one "thing." When taken as a whole, these components illuminate a business in great detail, allowing for better decision-making. In order to improve performance, business intelligence has grown to include more processes and activities in recent years. Among these procedures are:

- **Data mining:** Data mining is the process of discovering patterns in large datasets through the application of ML, statistics, and databases.

- **Reporting:** Data analysis is shared with stakeholders so they may make decisions and form conclusions.

- **Performance metrics and benchmarking:** evaluating performance versus objectives by comparing recent and previous data, usually with the use of Personalised dashboards

- **Descriptive analytics:** Analyzing early data to determine what transpired

- **Querying:** Asking queries specific to the data and using BI to extract the answers from the data sets

- **Statistical analysis:** deducing the cause and effect of this pattern from descriptive analytics and other statistical investigations of the data

- **Data visualization:** Data visualization involves transforming data analysis into visual representations such as charts, graphs, and histograms in order to make data consumption easier.

- **Visual analysis:** making stories using images to quickly communicate concepts and keep the analysis moving

- **Data preparation:** Data preparation is collecting information from many sources, determining its dimensions and measurements, and preparing it for analysis.

Although data analytics and business analytics are part of business intelligence, they are utilised in conjunction with other aspects of the process. With BI, users may draw conclusions from data analysis. Data scientists sift through data in search of patterns and trends by applying advanced statistical methods and predictive analytics.

The issue that data analytics seeks to answer is, "What may occur subsequent to this and what caused this to occur?" With the help of business intelligence, these models and algorithms are translated into a language that can be put to use. In the IT language of Gartner, "business analytics includes statistics, data mining, predictive analytics, and applied analytics." In conclusion, business analytics is an integral part of organizations' larger business intelligence strategy.

In order to aid in planning or decision-making, BI provides rapid analysis in response to specific queries. Nevertheless, companies might improve iteration and follow-up questions with analytics techniques. A linear approach is inappropriate for business analytics since solving one problem is likely to lead to other issues and iterations. Data access, exploration, discovery, and information exchange create a cycle in this process. This is called the "cycle of analytics," a term that has recently been used to describe how businesses use analytics to address changing questions and needs.

Business Intelligence (BI) can be defined as the methods, tools and procedures that enable an organization to gather, process, analyze and report business information to support the decision-making process. It is the process of analyzing data, meaning the conversion of raw data and its use in business intelligence to reveal patterns and track performance by providing a framework to analyze everything. BI systems are generally composed of some elements for instance data warehouse system, executive dashboards, reporting tools and data visualization tools amongst others in offering an all-inclusive outlook of an organization. As the name suggests, the use of BI enhances competitive advantage, productivity and overall organizational ability to meet market forces and customer needs.

1.3 Evolution of Business Intelligence

BI solutions have mostly concentrated on gathering a significant amount of data and providing business analysts with reports on it. But during the past 20 years, BI has seen some significant changes.

Therefore, modern BI technologies have become increasingly valuable. Today, business intelligence is driven by advances in AI and machine learning. In addition, several examples can be used to illustrate the growth and development of the business intelligence era

It must be borne in mind that the term "business intelligence" (BI) has been abused for quite some time, and that this definition is from relatively recently. Formally known as "traditional business intelligence," the term emerged in the 1960s as a means for companies to share data with one another. The term "business intelligence" was initially introduced in 1989 in conjunction with computers models for decision-making. These programs underwent additional development, turning data into insights, until they became a specialized offering from business intelligence teams via IT-dependent service solutions. There will be more to come from this essay as it introduces the reader to BI.

Traditional era of business intelligence

In the early days of business intelligence, the IT department oversaw all of the company's data. They started offering a way to combine data from multiple sources into a single database. The procedure is called ETL, which stands for extraction, transformation, and loading. On top of that, the company may assess the information. The enquiries would be handled by the IT staff on behalf of the clients.

Business owners then received a report from IT personnel. Days or weeks may pass throughout the procedure, depending on how skilled the crew is. The consumers also found the procedure to be ineffective and too lengthy.

Business intelligence's self-service Golden Age

The era of rapid computer development also saw a rise in the creation of business intelligence solutions. With the help of BI technologies, business users can access any and all data. With the help of these tools, business analysts can perform ad hoc analyses on the data sources (Adebunmi Okechukwu Adewusi et al., 2024)the proliferation of Big Data has revolutionized the way organizations gather, process, and utilize information for strategic decision-making. This paper provides a comprehensive overview of the evolving role of Business Intelligence (BI.

They facilitate the rapid discovery of patterns in massive data sets by data analysts. Instead of the traditional rows and columns used by more traditional data presentation tools, they make use of visual representations of the data through charts and graphics.

Consequently, business analysts' transition from the previous Extract, Transform, and Load (ETL) system was highly effective. Businesses may now analyses data more quickly because to this transformation. The business will be able to make a choice more quickly. Thus, they can overtake and contend with the rivals.

It would appear that modern BI is developing into both an art and a science all at once. Staff members, stakeholders, and the company overall may have an easier time seeing significant signs. It was only natural for many BI vendors to start including these state-of-the-art data analyses and visualization capabilities in their products.

Also, everyone who used the platform wanted data visualization and analysis to have enterprise-level features and data services. Reporting capabilities, data management, and security are all part of it.

Augmented analytics era of business intelligence

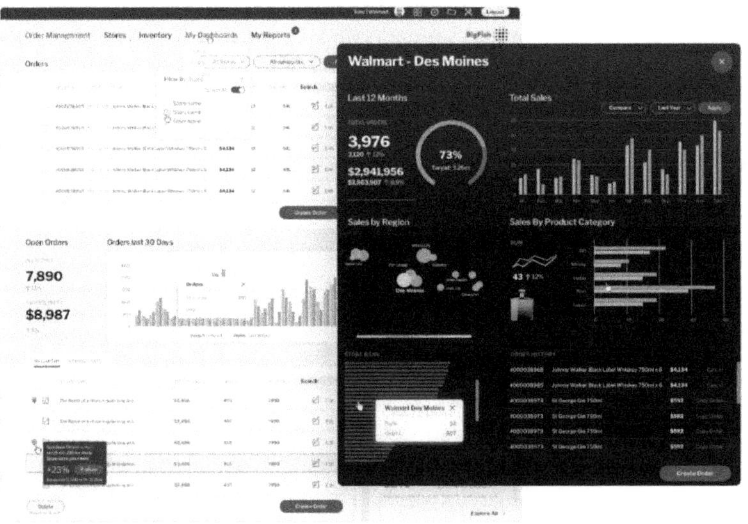

Figure 1.3: the above figure shows the different BI tools dashboard

Source: *- (Tableau, 2020)*

Data scientists nowadays are responsible for extracting actionable insights from massive datasets. At this point in time, we call it the "augmented analytics era."

This trend, known as augmented analytics, is causing the industry to move away from self-service analytics and towards automation.

The most recent business intelligence (BI) solutions are made to assist companies in producing insights that will enable them to operate better. By automating the process of producing insights, augmented analytics might lessen the need for data scientists.

Furthermore, more and more complex analytics technologies, such as machine learning, are being used. Businesses can automate the process of creating massive data sets and detect patterns and anomalies in the data with the help of tools provided by several popular Business Intelligence (BI) systems. Some BI solutions go so far as to perform analytics and provide insights all by themselves.

The history of Business Intelligence can be dated back to the year 1960s where first decision support systems where developed in organizations. These systems comprised primarily of data warehousing, which were designed to aggregate structured data from transactional databases and to provide reports for tactical utilization in the planning process. In the 1980s and early 1990s, BI tools progressed to data warehouses and OLAP technologies that stored vast amounts of information and made it feasible to make several dimensions of analysis. During this period, it was all about consolidating data to make history available for analysis to improve organizational decision-making.

In the 2000s, BI tools evolved and people had self-service analytics which enabled them to work on the data by themselves without so much dependence on IT personnel. Due to the availability of different kinds of data visualization tools and interactive dashboards, users found it convenient to analyse large amounts of data. The BI market has evolved and changed in the last decade moving to cloud solutions and augmented analytics that use AI & ML. This evolution is as a result of increasing need for faster analysis of data, improved models and ability to make better decisions within a competitive business environment.

1.4 Emergence of Cloud Computing

In 1963, MIT received $2 million for Project MAC from the Defence Advanced Research Projects Agency (DARPA). In order to be eligible for the funding, MIT had to develop a system that would allow a "computer to be used by two or more people, simultaneously." In this case, what we now often call "cloud computing" had its origins in one of those massive, archaic machines that stored data on reels of magnetic tape. It served as a basic cloud when used by two or three people. This situation was initially referred to as "virtualization," although the meaning of the term was later expanded (Sinaga-Bulgamin, 2022).

The Advanced Research Projects Agency Network, or ARPANET, was established in 1969 with the help of J. C. R. Licklider and was a "very" primitive version of the Internet. The idea of the "Intergalactic Computer Network," put forth by the psychologist and computer scientist JCR, also referred to as "Lick," would enable the whole population to be linked via computers and access data regardless of their physical location. (What would the appearance of such an unrealistic and financially impossible ideal of the future be like?) The Intergalactic Computer Network, or the internet, is necessary to reach the cloud.

The word "virtualization" has been used since the 1970s, and it now describes creating a computer simulation complete with an operating system that can run real-world programs. With the advent of "virtual" private networks for rent, the concept of virtualization evolved in tandem with the internet. The current architecture for cloud computing has its roots in the widespread adoption of virtual computers in the 1990s.

- **Cloud Computing in the Late 1990s**

  Cloud computing originally stood for the gap that existed between service providers and their customers. In 1997, Emory University professor Ramnath Chellapa characterised cloud computing as a new "computing paradigm, where the boundaries of computing will be determined by economic rationale, rather than technical limits alone." Using this rather lengthy description, we may trace the evolution of the cloud (TOADER, 2022).

The cloud's popularity skyrocketed as companies became more informed about its features and advantages. In 1999, Salesforce sprang to fame as a prime example of a successful cloud computing platform. The idea of selling software to consumers over the Internet was born out of their use of it. The program can be accessed and downloaded by anybody with an Internet connection. Businesses can purchase the software on demand and at a reasonable price without leaving their offices.

- **Cloud Computing in the Early 2000s**

  Online shopping services were introduced by Amazon in 2002. Though it was common practice at the time, this was the first major corporation to think about using only 10% of its capacity as a problem that needed fixing. With the help of the cloud computing concept, they were able to maximize the potential of their machine. Their strategy was quickly followed by other large businesses.

  In 2006, Amazon launched Amazon Web Services, a platform that offers various online services to third-party companies and customers. Among the many cloud-based services provided by Amazon Mechanical Turk—a website of Amazon Web Services—are storage, processing, and "human intelligence" among many others. Users can rent virtual computers and run their own applications on the Elastic Compute Cloud (EC2), another website provided by Amazon Web Services.

  Google also introduced its Google Docs service that same year. Two products, Google Spreadsheets and Writely, were sequentially derived from Google Docs. One web-based word processor that Google just acquired is Writely. Tenants may use it to create, modify, and save documents, as well as import them onto blogging platforms. They are also compatible with Microsoft Word because they are in the Word format. After acquiring 2Web Technologies in the middle of 2005, Google Spreadsheets became a web-based program that lets users create, modify, and share spreadsheets online. This one is built using Ajax and is compatible with Microsoft Excel. It is possible to save the spreadsheets as HTML files, which is relevant to the data mentioned above.

In 2007, IBM, Google, and other universities created a server farm for research assignments that needed massive datasets and fast CPUs. At the outset, the University of Washington was the pioneering institution to collaborate with Google and IBM. After it came the likes of Stanford, MIT, Carnegie Mellon, and the University of Maryland, among others. In a short amount of time, academic institutions understood that they could conduct computer experiments more cheaply and rapidly with the help of IBM and Google. Due to the fact that Google and IBM were both able to focus on areas of interest to them, the agreement was mutually beneficial. Along with the "binge-watching" trend, Netflix launched its cloud-based streaming service in 2007.

When it came to private cloud distribution, Eucalyptus had the first platform that worked with the AWS API, and that was in 2008. In the same year, Open Nebula, a NASA project, released the first open-source software for private and hybrid cloud deployment. Its most innovative features were mostly designed to meet the needs of huge organizations.

- **2010 and Beyond**

Private clouds were unpopular and still in their early stages when they were launched in 2008. Concerns about the security of public clouds prompted many to switch to private ones. Amazon Web Services (AWS), Microsoft Azure, and OpenStack's private clouds were mostly operational by 2010. (Additionally, in 2010, OpenStack launched a popular, free, open-source, DIY cloud that anybody could use.)

Hybrid clouds were initially suggested in 2011. Both public and private clouds need to be reasonably interoperable and able to transfer workloads amongst each other. Public clouds offer resources and storage that few firms had at the time, but many are interested in using this because of the potential benefits it could bring.

As a component of their cultural thinking program Smarter Planet, IBM debuted their IBM Smart Cloud architecture in 2011. Apple then launched iCloud to facilitate the storage of additional personal data, including images, videos, music, and more. Microsoft also began airing TV commercials this year touting the cloud and its many useful features, such as the ability to easily save videos and photos online.

Oracle introduced the Oracle Cloud in 2012, which encompassed the three pillars of any successful enterprise: Software as a Service (SaaS), Platform as a Service (PaaS), and Infrastructure as a Service (IaaS). These "basics" quickly became the standard, even though some public clouds focused on delivering only one service while others offered all of them. The SaaS model became very popular.

In 2012, Cloud Bolt was founded. In order to assist organisations construct, implement, and manage both private and public clouds, this business is credited for creating a hybrid cloud management platform. Public and private cloud interoperability issues were fixed by them.

Businesses began using software as a service (SaaS) providers for specialised functions, such as human resources, supply chain management, and customer relations management. This led to the emergence of multi-clouds. It wasn't until 2013 or 2014 that this started to become popular. While many still rely on SaaS providers, a new idea has evolved around utilising many clouds, each with its own set of advantages. A component of this outlook is the desire to avoid being compelled to use a specific cloud because of "interoperability issues.".

By 2014, the core characteristics of cloud computing had been defined, and security had become an important consideration. The fast expansion of cloud security as a service is a direct result of the importance it has on consumers. The security of cloud computing has come a long way in the past several years, and it is now competitive with more traditional forms of IT protection. Theft, data leaks, and accidental deletion are all part of this. But security is, and probably will continue to be, the biggest concern for most cloud users.

These days, application developers are among the main consumers of cloud services. Developer-driven cloud computing started to replace developer-friendly cloud computing in 2016. Application developers started making the most of the cloud's resource capabilities. Many services aim to attract more clients by being developer-friendly. Understanding the necessity and the possibility for financial gain, cloud providers created (and are still creating) the tools that app developers want and desire.

- **Containers**

  Solaris containers were there since 2004, but they were severely constrained and could only be used with certain computer systems. Prior to Docker's 2013 release of a very practical container, these technologies did not achieve widespread adoption. It is not a coincidence that the usage of Docker and containers increased simultaneously.

  Hundreds of tools that have been there for a while were updated and put to use in 2017 to make container operations easier. One of these was Kubernetes, an open-source software platform developed and released by Google in 2014. Kubernetes is a container orchestration platform that can automate the deployment, scaling, and administration of an application.

Cloud computing as a disruptive technology was successfully introduced in the early 2000s as a new method of storing, managing and accessing data and applications. It evolved from ideas such as utility computing and virtualization, which called for computing services to be made available over the internet on-demand. The emergence of such services as Amazon Web Services (AWS) in 2006, that brought the focused IaaS solutions for scale. This indicates that in the internet a business is in a position to acquire computer power, storage and application as a service without necessarily undertaking large installations at its premises. This change has brought flexibility, cost saving and integration and new opportunities for a new calculations, artificial intelligence, and home/remote working in all sectors.

1.5 Defining Cloud BI

BI (Business Intelligence) applications based on the cloud concept give business data a level of openness it has never experienced before. Their simplicity, digestibility, effectiveness, and flexibility make them suitable for organisations seeking quicker and better ways of exploring and interpreting data. (Al-Aqrabi et al., 2015).

The time has come to think about how cloud business intelligence may help your team and your company.

What is cloud BI?

Applications for cloud business intelligence (BI) are centralised in a virtual setting that includes the Internet. They are usually used to give organisations access to BI-related data, such as afternoon, KPI, and business analytics data, among other things.

Cloud products and services are sought after by businesses, such as CRM systems for automating sales, online document sharing and storage platforms like Dropbox and Box, and help desk tools like Zendesk and User Voice. The adoption of cloud space and cloud solutions' flexibility by business intelligence software is another indication of this trend.

Cloud BI vs. traditional BI

Getting the appropriate information at the right time in the right place is the goal of business intelligence, and cloud computing makes BI tools and applications lightweight and flexible.

To help you identify the best options for your business, the following is a breakdown of the differences between cloud BI and traditional BI:

- **Cost efficiency:**

   Let's consider the following benefits with cost being the first; a cloud solution is affordable. While, BI has turned into an application service, requiring large initial investments in hardware, software and BI infrastructures, Cloud BI charges its clients based on a per user per month basis.

   Thus, small businesses are left with little costs to incur before acquiring these resources in this approach. In addition, it also decreases or even eliminates costs associated with continuous maintenance, modification and employing of IT personnel.

   Therefore, avoiding a cloud-based business intelligence tool could prove to be quite expensive in the long term. It allows you to more efficiently deploy resources in other areas of your company.

- **Ease of use:**

   Conventional business intelligence is sometimes complicated and difficult to utilize due to its age. The technology behind cloud BI software, on the other hand, is more advanced. This makes it possible for developers to produce software that is easy to use, with simplified features and intuitive interfaces.

With cloud BI software, even non-technical people can now easily create reports and visualizations, whereas previously, business intelligence could only be achieved by writing complex queries and reports.

For example, you don't need to write any code to generate stunning data visualizations using Power Metrics. Our business intelligence platform makes it simple for everyone in your team to exchange data and develop sound plans.

- **Deployment speed:**

  Since cloud apps don't require any additional hardware or software installs, they are incredibly easy to deploy. You may avoid the time-consuming procedures of purchasing and configuring physical servers, as well as configuring software and networks, by using these business intelligence applications.

  In a fraction of the time required to establish a standard BI system, you can have your BI system operational. Furthermore, the offered cloud solutions are adaptable as they let you quickly change your BI environment. This allows your company to swiftly adjust to changes in the market or within the company.

- **Scalability and elasticity:**

  In the course of conducting business, the kind of data analysis required by your company will eventually change. Depending on the requirements, cloud BI solutions may be swiftly improved or reduced in size. As a result, you have flexibility that is not possible with traditional BIS that has set capacity.

  This degree of scalability may be attained, allowing for a more thorough evaluation of BI capabilities in relation to business requirements than more inflexible configurations can provide.

- **Accessibility:**

  With a variety of on-demand access options to precise data and information, Cloud BI helps your team to remain consistent. Users can monitor business intelligence dashboards and create reports as needed while they are at work, remotely, or on the go. This enables them to make timely and informed decisions, which is crucial for maximizing opportunities.

Cloud BI integration capabilities

Another area in which cloud BI solutions excel is integration. It is crucial that they have the ability to connect to multiple data sources, particularly when the company is likely to work with multiple datasets from different platforms.

- **Databases:**

  These days, standard databases in the BI industry include traditional relational business intelligence databases like MySQL, Oracle, and SQL Server. They include well-structured data that is important for the operation of your business and even for one-time transactions. Files remain the solid, dependable and dependable molds which have helped businesses for so many years.

  Conversely, the contemporary NoSQL DBs, like MongoDB or Cassandra, provide rather freewheeling treatment of tremendous amounts of unorganized or semiformal data. Organizations require a flexible BI solution to address the integration of multiple data categories and the ability to quickly grow.

  Incorporation of cloud BI with these databases enables one to access large data storage for enhanced and detailed analysis of reports needed in decision making.

- **Cloud storage:**

  Knowledge is the source of power. As a result, cloud storage services like Amazon S3, Google Cloud Storage, and Microsoft Azure Blob Storage are very advantageous to businesses.

  They provide safe, scalable spaces for companies to store enormous volumes of unstructured data, such as transaction logs and customer interactions. You can access and analyses data at any time using cloud business intelligence solutions, turning unstructured data into plans that can be put into action.

- **SaaS applications:**

  Applications such as Slack, HubSpot, and Salesforce are more than simply company operations management tools. They are also excellent sources of data. Because Cloud BI can interface with various apps, you can use customer and operational data.

- **Spreadsheets and files:**

  Spreadsheets and common file formats like Excel and CSV are still used by many organisations. This is most likely due to the fact that these business intelligence repertoires are easily available, highly adaptable, and user-friendly.

- **APIs:**

  Application Programming Interfaces, or APIs, act as links between various software systems, facilitating smooth data sharing and communication. Real-time data from outside sources may be retrieved by cloud business intelligence solutions to provide you with a wider variety of information.

- **IoT Devices:**

  Because of the Internet of Things (IoT), corporate intelligence and data collection have entered a new era. These days, a lot of devices are providing real-time streams of data continuously.

The data from these devices is crucial for industries that rely on timely information for operational efficiency and predictive maintenance. For example, industrial companies use IoT data to streamline operations and reduce disruptions.

Cloud Business Intelligence (Cloud BI) can be described as a delivery model of Business Intelligence tools and services with a cloud delivery model where organizations can access, analyze and share data to anyone regardless of their location. Traditional BI is strengthened by the advantages of the cloud computing approach, which also saves companies from intricate setups on premises. Some of the cloud BI solutions and services include data integration, stylish and lively dashboards, real-time reports, elaborate analytics AI, and ML. With the help of cloud technology one can make decisions based on data more quickly, improve teamwork, and save money to maintain physical server rooms.

1.6 Advantages of Cloud BI over Traditional BI

The contemporary method of connecting and evaluating data from the cloud to obtain insightful information for decision-making is known as cloud business intelligence, or cloud BI. By linking disparate data sources, including financial, sales, and customer information, organisations may uncover significant patterns and anticipate future trends. Users can take automated activities based on the insightful information provided by cloud BI solutions.(Sun et al., 2018).

Businesses seeking quicker and more effective ways to analyses and visualise data will find cloud-based business intelligence solutions perfect due to their user-friendly design, quick deployment, scalability, and accessibility.

Cloud business intelligence solutions gather, organize, switch, and convert data using a number of sophisticated techniques. Cloud BI applications link to pre-existing data sources, such databases and spreadsheets, and process the data using a variety of techniques, including statistical models and algorithms. Cloud business intelligence is prepared to analyses this data and find insights. These insights may be shared with other users, shown in data visualizations like bar charts or histograms, and utilised to guide choices that improve company performance (Sandhu, 2022)the amount of data available is increasing day by day. However, excessive amounts of data create great challenges for users. Meanwhile, cloud computing services provide a powerful environment to store large volumes of data. They eliminate various requirements, such as dedicated space and maintenance of expensive computer hardware and software. Handling big data is a time-consuming task that requires large computational clusters to ensure successful data storage and processing. In this work, the definition, classification, and characteristics of big data are discussed, along with various cloud services, such as Microsoft Azure, Google Cloud, Amazon Web Services, International Business Machine cloud, Hortonworks, and MapR. A comparative analysis of various cloud-based big data frameworks is also performed. Various research challenges are defined in terms of distributed database storage, data security, heterogeneity, and data visualization.",”author”:[{“drop ping-particle”:””,”family”:”Sandhu”,”given”:”Amanpreet   Kaur”,”non-dropping-particle”:””,”parse-names”:false,”suffix”:””}],”container-title”:”Big Data Mining and Analytics”,”id”:”ITEM-1”,”issued”:{“date-parts”:[[“2022”]]},”title”:”Big   Data   with   Cloud   Computing: Discussions   and   Challenges”,”type”:”article-journal”},”uris”:[“http:// www.mendeley.com/documents/?uuid=bc4f323f-7ace-42a5-ae72-00fac768bd80”,”http://www.mendeley.com/ documents/?uuid=bb973e24-0c93-4695-85e6-bdc3662f3bbc”]}],”men deley”:{“formattedCitation”:”(Sandhu, 2022.

Take a Examine how cloud business intelligence may improve overall business performance and increase your team's understanding.

Cloud Business Intelligence Solutions' Main Advantages

As mentioned above Cloud Business Intelligence solutions present many benefits which can change the functioning of all companies.

The above solutions offer affordability, increased data availability, better cooperation, quicker deployment, innovative security, compatibility, and instantaneous analysis. Using the benefits listed above, companies will be able to analyze things through the usage of data, and this makes things easier for them when they are competing in an industry.

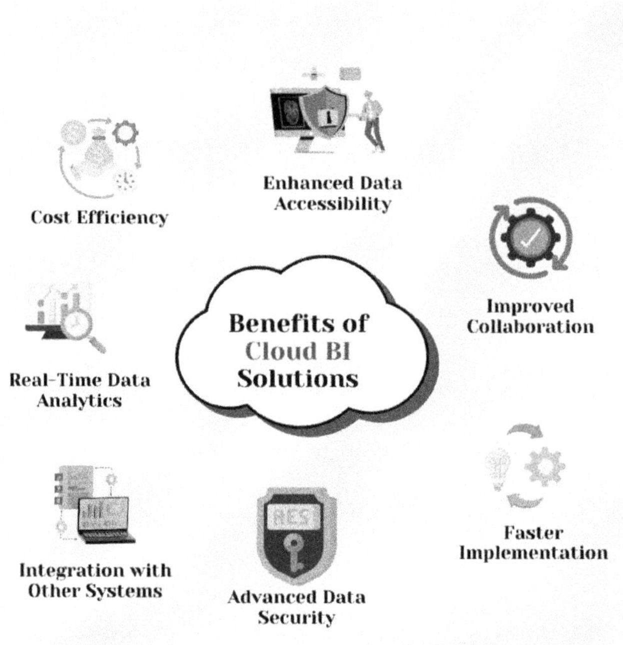

Figure 1.4: Key Benefits of Cloud Business Intelligence Solutions

Source: - *(Signatech, 2022)*

Scalability and Cost Effectiveness:

Cost saving is one of the major advantages of implementing cloud-based business intelligence solutions. In a cloud computing environment, one cannot have to invest significant amounts of money that are needed to support the establishment of local infrastructures. Rather, they make payments based on their usage hence a better mode of operating an organization's expenses. Also, cloud solutions have flexibility inherent in them which allows a firm to quickly add or remove more resource capacity depending on needs without having to profoundly alter the underlying infrastructure to cater for this.

- **Reduced Capital Expenditure:** Cloud computing allows businesses to avoid the expensive need to buy software and hardware.

- **Pay-as-You-Go:** Most of the Cloud BI solutions are available with subscription plan, which enables the business to pay closely with the usage.

- **Elastic Scalability:** Organizations can adjust their BI resources accordingly to the current requirement without much investment on additional infrastructure.

Improved Accessibility of Data:

Cloud Business Intelligence solutions give unprecedented levels of access to data. The beauty of using data and BI tools is that users can retrieve them from anywhere in the world with internet connectivity. This level of accessibility guarantees entrepreneurs real-time access to some of the most important business indicators thus making it easy for them to take rational decisions at any given time.

- **Anytime, Anywhere Access:** Managers and other decision-makers can then access or receive information from any terminal whether it is a desktop, tablet or a smartphone.

- **Facilitation of Remote Work:** As a result of cloud BI, different teams across organizations can work on the same BI environment making them to be up to date even they are working from different parts of the world.

- **Centralized Data Repository:** One of the primary advantages of Cloud BI solutions is that usually the data is stored in a single place and is easier to extract across the sources.

Enhanced Cooperation and Efficiency:

In the cloud environment, there is enhancement of cooperation among the members of a specific team. In organization BI solutions, there are usually social capabilities that enable many users to use the same cloud BI in their respective areas then share the discovered data, ideas, and reports. It also optimises efficiency and makes it possible to have all the stakeholders in a project at the same page.

- **Real-Time Collaboration:** Team members can edit, revise or analyze the data sets at the same time as other members of the team, which means that discoveries are often discussed as they are made.

- **Shared Dashboards and Reports:** Through Cloud BI, the creation of dashboards and sharing of reports are possible thus enabling wide sharing of information in an organization.

- **Enhanced Communication:** The base allows for direct communication and integration with other related applications (such as Slack or Microsoft Teams) to help with collaboration and to optimize workflows.

Faster Implementation and Updates:

Implementing of traditional BI solutions might take a lot of time and efforts. Cloud Business Intelligence solutions, however, are faster when it comes to implementation. Companies are also able to easily deploy their BI tools and get value out of their data very fast. In addition, the updates and maintenance is conducted by the cloud providers which means that the system is updated with latest features and with the security patches when required.

- **Quick Deployment:** In comparison to the traditional systems, Cloud BI solutions are in a position to be deployed in days or weeks.

- **Automatic Updates:** It is the responsibility of cloud providers to take the initiative of updating the cloud, making sure everyone has the updated cloud complete with the latest features and security update.

- **Reduced Downtime:** This means few waves made to the business since routine maintenance and updates are done by the provider.

Advanced Data Security:

Security is, of course, always a major concern when it comes to any business's performance. To protect sensitive company data, cloud business intelligence systems are protected by robust, well-known security mechanisms. Major cloud service providers provide high level security measures like data encryption, PAM authentication, and security checkups to keep your data safe.

- **Data Encryption:** Data both while idle and while in transport are also encrypted to ensure their security.

- **Multi-Factor Authentication:** Improvement in the identities' verification mechanism avoids access of unauthorized individuals.

- **Regular Security Audits:** It is necessary to mention that performing constant checks on that and having regular security audits will reveal possible threats.

Smooth Interaction with Different Systems:

Cloud business intelligence (BI) solutions are designed to be readily connected with other data sources and business processes. A CIO can get a comprehensive company view by using cloud BI solutions that can collect data from multiple sources, such as marketing automation platforms, CRMs, and ERPs. Your BI investment is increased by this integration capabilities, increasing its value.

- **Thorough Data Analysis:** Combining the data extracted from different sources is basically more suitable and beneficial as part of improving the analysis.

- **Unified Data View:** Single source of truth could be whereby businesses integrate more than one data source from different platforms into one BI solution.

- **Streamlined Workflows:** Interaction with other business systems makes the work process easier and more effective.

Real-Time Data Analytics:

One of the most significant strategic advantages that any company may have is real-time data analysis. Cloud business intelligence solutions are real-time because they enable businesses to process data as it occurs within the company. It makes a lot of sense for businesses like retail, banking, and healthcare where quick decisions based on current data are essential.

- **Instant Context:** Real-time analytics enable companies to act quickly on trends and issues as they develop.

- **Enhanced Decision-Making:** Availability of data is useful enhances the decision-making process by minimizing delays.

- **Competitive Advantage:** Organizations can be able to be competitive and respond rapidly to the changing market trends through the use of real time data.

Adaptability and Personalization:

Cloud BI solutions are very flexible and scalable. The BI tools for small business can be customized in accordance with all demands thus individual specially developed dashboards and reports can help businesses adjust for their needs. This level of personalization makes sure that businesses receive the right information they need to act on.

- **Custom Dashboards:** Dashboards may be customized by users based on their own measurements and KPIs.

- **Adaptable Reports:** Reports can be modified to highlight the most important details for different stakeholders.

- **Scalable Solutions:** Cloud BI solutions are scalable and can be easily adjusted to elementary or more complex needs of the expanding business.

Reduced IT Dependency:

As a result, Cloud BI helps businesses decentralize their reporting and analyzing process by lowering the dependency on IT departments. Cloud BI tools are usually easy to use, long-range designed to be used by business end-users rather than IT specialists to create reports and analyze results. That fosters teamwork and brings more discipline to the decision-making system.

- **User-Friendly Interfaces:** Easy to use interfaces lets those who are not technically inclined fully enable and interact with BI tools.

- **Self-Service BI:** Data consumers can monitor and analyze information on their own without significant reliance on IT teams.

- **Faster Decision-Making:** Users are enabled to transform data into usable knowledge and take decisions there and then without having to consult the IT department.

Environmental Sustainability:

Cloud computing, as a rule, is less energy-intensive than on-premises data centers counterparts. Cloud BI solutions also help organizations minimize the negative impact on the environment because they help to decrease carbon footprint. Cloud providers also purchase advanced power optimizing technologies and renewable energy as a part of the commitment to green programs.

- **Energy Efficiency:** Concrete data processors are usually less energy effective than their equivalents within the clouds.

- **Reduced Carbon Footprint:** At the same time, through the utilization of cloud services, companies reduce the amounts of energy and emissions they use.

- **Sustainable Practices:** Big cloud providers have pledged to use only renewable energy and operate sustainably.

Cloud Business Intelligence or Cloud BI is a much better solution than traditional Business Intelligence systems for several reasons – firstly, Cloud BI is scalable and easily accessible; secondly, Cloud BI is much cheaper. While traditional BI solutions require upfront capital expenses for hardware, software, and maintenance, Cloud BI works with monthly subscriptions, thus extending the possibility of carrying out the continuous advanced analytics to all types of businesses. Its web interface allows users to search for and process data from any location and avoid one-on-one teamwork and remote work. Cloud BI also has easy definitions of data and provides real-time information so that business can make right decisions quickly. However, with the help of cloud providers, updates and security are presented automatically and are enhanced continually without the need to involve IT professionals.

1.7 Key Components of Cloud BI Solutions

Cloud Business Intelligence (Cloud BI) solutions include multiple components that define how businesses can gather, store, analyse, and leverage data. These components make sure that the decision making based on data is made easy and is available in an organization. Knowledge of these three areas allows for deploying integrated BI solutions and optimizing the possibilities of cloud technology (Balachandran & Prasad, 2017)without a doubt, two of the most important technologies to enter the mainstream IT industry in recent years. Surprisingly, the two technologies are coming together to deliver powerful results and benefits for businesses. Cloud computing is already changing the way IT services are provided by so called cloud companies and how businesses and users interact with IT resources. Big Data is a data analysis methodology enables by recent advances in information and communications technology. However, big data analysis requires a huge amount of computing resources making adoption costs of big data technology is not affordable for many small to medium enterprises. In this paper, we outline the the benefits and challenges involved in deploying big data analytics through cloud computing. We argue that cloud computing can support the storage and computing requirements of big data analytics. We discuss how the consolidation of these two dominant technologies can enhance the process of big data mining enabling businesses to improve decision-making processes. We also highlight the issues and risks that should be addressed when using a so called CLaaS, cloud-based service model.","author":[{"dropping-particle":"","family":"Balachand ran","given":"Bala M.","non-dropping-particle":"","parse-names":false ,"suffix":""},{"dropping-particle":"","family":"Prasad","given":"Shivika ","non-dropping-particle":"","parse-names":false,"suffix":""}],"contain er-title":"Procedia Computer Science","id":"ITEM-1","issued":{"date-parts":[["2017"]]},"title":"Challenges and Benefits of Deploying Big Data Analytics in the Cloud for Business Intelligence","type":"paper-conference"},"uris":["http://www.mendeley.com/documents/?uuid=52b49642-c8d4-4229-8caa-14bc61977e87","http://www.mendeley.com/documents/?uuid=a87c3d1a-5062-4073-ab37-e1 37a2461a5b"]}],"mendeley":{"formattedCitation":"(Balachandran   & Prasad, 2017.

1. **Data Sources and Integration:** Data connection is another feature of Cloud BI that encompasses the capacity to share with other data sources like databases, spreadsheets, cloud repositories, and third-party solutions. The major advantages of Cloud BI platforms are the tools on data integration, which are efficient for importing and consolidating the data then move on the data warehouse. It also guarantees one truth on data in different departments hence reducing variations. ETL, for example, is an aid for process standardization and data preparation for analysis and real-time data connectors.

2. **Data Storage and Management:** Cloud BI solutions make use of elastic cloud storage to handle the fixed and variable structured and unstructured data. There are several data storage technologies that are used as hybrid data storage structures between data lakes and data warehouses, including Amazon Redshift, Google Big

Query, and Snowflake. They also support distributed storage so that availability is high and the ability to recover from disaster is quick with redundancy. Cloud BI platforms also integrate the data governance feature, and role-based access control (RBAC) and encryption to ensure information security.

3. **Data Analytics and Processing:** Cloud BI solutions are characterized by superior data analytics and processing systems. Several Cloud BI tools that use machine learning (ML) and artificial intelligence (AI) to automate the process of analysis, identification of anomalous trends, and trend forecasting also use these platforms, which use cloud-based HPC resources to execute analytical algorithms for statistical analysis, predictive modelling, and trend analysis. Current data processing characteristics allow businesses to have access to the timeliest information for timely decision-making.

4. **Data Visualization and Reporting Tools:** The visualization and the reporting of data, in general, involve computer programs as they are crucial in data analysis. Cloud BI solutions offer unique dashboards, graphic, and drag-and-drop report tools that allow users to analyse data regardless of their experience in data analysis. Tableau, Power BI and Looker enables the organization to prepare stunning visual reports wherein the raw data is translated and made easy to digest. Real-time dashboards also add to effective KPI display and tracking of business metrics.

5. **Collaboration and Accessibility:** Collaboration and accessibility are vital aspects of Cloud BI solutions, making data-driven insights available to all stakeholders within an organization. Cloud-based platforms allow multiple users to access and work on data simultaneously, regardless of their location. Features like shared dashboards, version control, and role-based permissions ensure secure collaboration while maintaining data integrity. This accessibility promotes a data-driven culture where decision-makers at all levels can leverage insights to drive business growth.

1.8 Overview of Industries Benefiting from Cloud BI

Cloud Business Intelligence or Cloud BI has emerged as an enhanced method of using BI for embracing the data-driven approach to decision-making within industries by ensuring affordable, real-time, and flexible data usage. Cloud BI has now been adopted by different industries to help with effectiveness and effectiveness in performance, customer relations and innovation through the use of analytical data (Kasem & E. Hassanein, 2014).

1. **Retail and E-commerce:** These days, a number of businesses, like retail and e-commerce, use Cloud BI to assess user preferences, sales, and inventory. With help of purchasing data, it is possible to understand the client's needs and behaviour, tailor the marketing message, manage the inventory, and anticipate the future demand. Cloud BI also assists retailers to track multi-channel sales and make right decisions which will lead to improved profitability.

2. **Healthcare:** The application of Cloud BI in health care include: management of huge patient data, enhanced service delivery and support for research. HIPAA for example allows providers to examine the results of patient care, manage business operations and confirm with the requirements of healthcare data privacy law. Cloud BI tools also help in making predictive analysis so that probable diseases can be detected and the utilization of resources in different hospitals and clinics can be enhanced.

3. **Financial Services:** Cloud BI helps the financial services industry gain insights on market trends, risk management and the customers. Some of the Cloud BI applications are fraud investigation, credit risk evaluation, and portfolio management in many banking institutions such as insurance firms and investment companies. Real-time data processing improves the decision-making process and preserves financial compliance with the help of secure data management tools.

4. **Manufacturing and Supply Chain Management:** It has been identified that Cloud BI is used by manufacturers for optimizing production, as well as having quality assurance and supply chain transparency. Cloud BI solutions assist in the monitoring of the equipment usage, identification of constraints within the manufacturing process, as well as the predicting of when equipment might require maintenance in future. Cloud BI provides a clear insight on the supply chain, including the flow of products and their suppliers, reducing costs and increasing efficiency.

5. **Education:** Cloud BI is adopted in the education sector for tracking students' results and performance, and also making changes to the institutional operations. In the case of education, data has the potential to be used in helping identify learning loss, rate of learning and even designing learning experiences. Cloud BI also plays key roles in the resource planning, enrolment forecasting and faculty performance reporting.

6. **Marketing and Advertising:** Marketing and advertising organizations get value from Cloud BI in the form of insights as to how their campaigns are faring, which audiences are being targeted, and how customers are interacting with their marketing messages. Marketing communication channels allow marketers to monitor the performance indicators and ROI on the cloud and realign the strategies. It is much easier to come up with highly effective advertising and marketing strategies when you are able to intensely scrutinize the big data.

Due to the adaptable, expandable, and affordable nature of Cloud BI, a great number of industries can take advantage of data insights and achieve continuous improvement.

1.9 Streamlining DAX Query Creation and Insights with Copilot

Copilot may be used to write and explain Data Analysis Expressions (DAX) queries in semantic models using the DAX query view in Power BI. There are various tools available in the DAX query view to help you use DAX queries as efficiently as possible. The public is currently able to preview this feature. To find out how Copilot can help you be as effective as possible with DAX queries, see Copilot's DAX query features.

In the public preview, Copilot can write and explain DAX queries in the DAX query view. An inline Fabric Copilot is included in the DAX query view to help write and explain DAX queries, which are still in public preview.

1.   **Run the DAX query prior to storing it.**

The Run button was previously inactive until Copilot was closed or the resultant DAX query was approved. You can now run the DAX query and choose whether to keep or discard it.

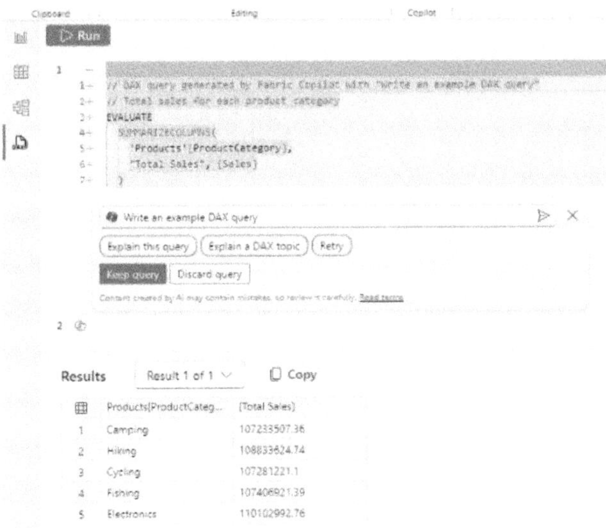

Figure 1.5: DAX query

Source: - *(MaggiesMSFT, 2024)*

2.   Create a conversational DAX query

Previously, if you typed more prompts, the DAX query that was generated was not taken into account, and you had to keep the DAX query, select it again, and then use Copilot again to adjust. Now, you can just type in more user prompts to make adjustments.

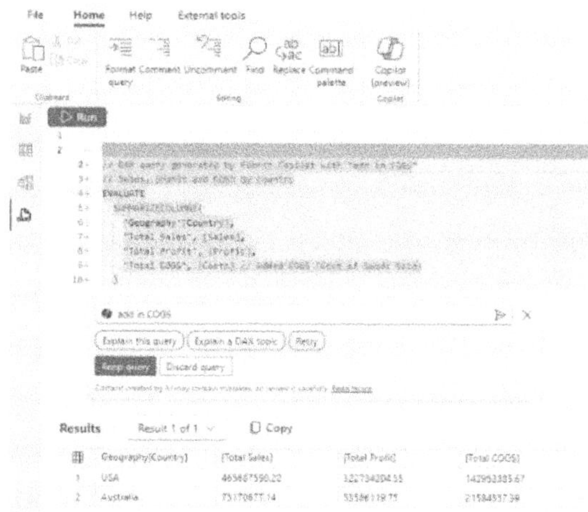

Figure 1.6: Build the DAX query

Source: - *(MaggiesMSFT, 2024)*

3.   Checks for syntax in the resulting DAX query

The generated DAX query was returned without a syntax check, but now it is checked and the prompt automatically retries once. If the retry is also unsuccessful, the generated DAX query is returned with a note indicating that there is a problem, allowing you to either fix the generated DAX query or rephrase your request:

" There are mistakes in this query. Try changing your request again or attempting to repair it yourself."

4.   Inspiring buttons to introduce user to Copilot

Before you entered a prompt, nothing happened. To rapidly see what you can accomplish with Copilot, click on any of these buttons.

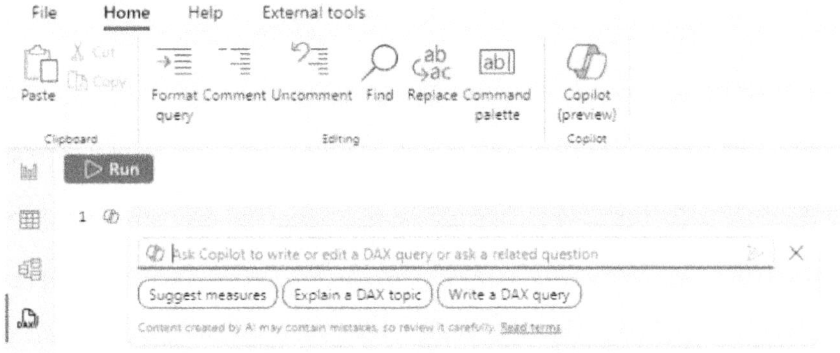

Figure 1.7: the generated DAX query

Source: - *(MaggiesMSFT, 2024)*

1.10 Chapter Summary

This chapter introduced the basics and the history of Business Intelligence (BI) and show how BI has grown from early decision support systems in the 1960s to today's cloud-based BI systems. Cloud BI solutions have become affordable, flexible and within the reach of modern businesses, and as a result, even such advanced tools as AI and ML became more accessible. Cloud BI allows quick decision making, improved collaboration and also integration of BI in different fields including retail, healthcare, finance and manufacturing sectors.

Along with collaborative tools, it also covered certain facets of Cloud BI solutions including data integration, storage, analysis, and visualization. Cloud BI has many benefits compared to the traditional BI solutions which makes it efficient for decision making: cost savings, real-time data, and flexibility. With Cloud BI becoming the trend for most organizations, firms are put at a vantage given the efficiency realized, proper resource utilization as well as quality insights.

Multiple Choice Questions (MCQs)

1. What is the primary focus of Business Intelligence (BI)?

    a.   Automating software development

    b.   Enhancing data-driven decision-making

    c.   Improving cloud infrastructure

    d.   Managing hardware systems

2. Which factor significantly contributed to the evolution of Business Intelligence?

    a.   Introduction of robotics

    b.   Emergence of big data

    c.   Decline in data storage needs

    d.   Elimination of cloud computing

3. What is a key feature of cloud computing that benefits Cloud BI?

    a.   Manual data processing

    b.   On-premises storage

    c.   Scalability and flexibility

    d.   Reduced data accuracy

4. How does Cloud BI differ from Traditional BI?

    a.   Cloud BI uses physical servers exclusively

    b.   Traditional BI requires no hardware investments

    c.   Cloud BI is accessible from anywhere via the internet

    d.   Traditional BI supports real-time analytics

5. **What is the primary advantage of Cloud BI over Traditional BI?**

   a. High maintenance costs

   b. Faster implementation and scalability

   c. Limited data storage options

   d. Dependency on local servers

6. **Which of the following is NOT a key component of Cloud BI solutions?**

   a. Data visualization tools

   b. Predictive analytics

   c. Physical hardware servers

   d. Data integration capabilities

7. **Which industry has seen significant benefits from Cloud BI?**

   a. Construction

   b. Manufacturing

   c. Healthcare

   d. All of the above

8. **What is a defining characteristic of Cloud BI?**

   a. Offline-only access

   b. Integration with cloud infrastructure

   c. Exclusively for large enterprises

   d. Minimal use of advanced analytics

9. **Why is Cloud BI becoming increasingly popular across industries?**

   a. High initial costs

   b. Improved accessibility and collaboration

   c. Dependence on physical servers

   d. Limited scalability options

10. **What summarizes the chapter on Cloud BI?**

    a. Cloud BI is exclusively for data storage.

    b. Cloud BI is a temporary trend in data management.

    c. Cloud BI combines traditional BI principles with the advantages of cloud computing.

    d. Cloud BI focuses only on hardware infrastructure.

Answer

1	2	3	4	5	6	7	8	9	10
b	b	c	c	b	c	d	b	b	a

Chapter 02

ARCHITECTING CLOUD BI SOLUTIONS

2.1 Chapter Overview

This chapter outlines the major issues which are crucial in creating effective and sustainable cloud-based BI systems. Starting with the scalability of data architectures in order to determine if solutions can economically scale with data growth. An analysis is made on the choice of the most suitable cloud service models such as IaaS, PaaS, and SaaS with regard to the needs of an organization is made. Some of the approaches advocated for making data integration smooth and for ensuring high quality and coherent data are presented as the best practices in cloud BI. As well as particularities of secure platforms and their compliance with regulations, techniques that allow attaining higher processing rates are discussed. Readers who have finished reading this chapter should be able to create and implement cloud BI systems that are safe, elastic, and highly available to satisfy the demands of modern businesses.

2.2 Designing Scalable Data Architectures

In the present information explosion age, companies and organisations are trying to control the digital equivalent of a wild, noisy horse running across the wide-open plains of cyberspace as they deal with an enormous inflow of data. Developing infrastructures that can grow with the data in a seamless way is a challenge, in addition to handling the massive volumes of data that are being received. In order to create scalable data structures that can handle increasing data quantities and future expansion, this article will examine best practices and guiding principles.

Understanding the Fundamentals of Scalable Data Architectures

The secret to data architecture scalability is creating systems that can easily adjust to increasing demands. There are two primary methods of scaling that are employed: vertical scaling and horizontal scaling.

In vertical scaling, present servers are improved, whereas in horizontal scaling, data and duties are dispersed among numerous servers. It's like putting together a team of superheroes. To create a diverse squad with a range of skills, you can use horizontal scaling to add more heroes and give them different tasks (Aron powers, 2023).

Conversely, vertical scaling is comparable to giving a single hero a significant boost in power, making them an extremely powerful person capable of handling increasingly challenging tasks. However, a fully scalable architecture's data modelling methodology ensures that the database structure may expand and alter without sacrificing speed.

Key Principles for Designing Scalable Data Architectures

Data must be partitioned and appropriate sharding techniques used in order to achieve scalability. Although they are similar ideas, partitioning and sharding are not the same. They both include dividing a big dataset into smaller, easier-to-manage chunks for better scalability and performance, but they may be implemented differently and are frequently used in distinct circumstances:

- **Partitioning** is a more general phrase that refers to the division of the data set into significant subgroups based on predetermined criteria. It is possible to organize these subsets, also known as partitions, based on many criteria, such as value, time, space, or any other criterion. Typically associated with conventional relational database systems, partitioning is a technique intended to improve query speed by allowing the database to operate with only the necessary scope of data.

- **Sharding**, is a more general phrase that includes dividing the data set into significant subgroups according to predetermined criteria. Some criteria, such as value, time, space, or any other requirement, can be used to organize these subsets, which are referred to as partitions. Usually associated with conventional relational database systems, partitioning allows the database to operate with only the necessary scope of data, improving query performance.

- Data replication and redundancies also carry a significant role in emitting the tolerance against the defective hardware and the data integrity. Distributed computing frameworks allow the option to exploit the fundamental notion of scalability, meaning parallel computation of vast volumes of data.

Best Practices for Building Scalable Data Pipelines

A constructed pipeline that outlines the procedures from data gathering to drawing conclusions is at the heart of every effective data utilisation strategy. This data symphony begins with the introduction of effective data intake, free from outside influence as each source creates its own unique melody. The music moves into the phases of transformation and purification from this opening chord. To illustrate a certain type of tale, it may be thought of as polishing the pure raw material until all undesirable imperfections are removed and a distinct unity of characteristics is brought out.

Choosing the appropriate tools for the Extract, Transform, Load (ETL) procedures is similar to picking great musicians who can perform every piece with perfection. How well the data journey goes depends on how effective and adaptable these technologies are. These ETL tools should be able to handle a variety of data types and structures with ease, just like talented musicians can adjust to many genres and styles. An all-purpose conductor like Tim extender can enter the scene at this point. Tim extender is the key that guarantees your data pipeline produces a beautiful and powerful performance, laying the groundwork for perceptive analytics and strategic decision-making, much like a maestro directing all the instruments to play in perfect harmony (Kujawski, n.d.).

Database Choices for Scalable Architectures

The relational versus NoSQL controversy frequently arises when selecting databases for scalable architectures. Relational databases, as you probably already know, use a tabular, organized style with rows and columns for data storage and keys to create associations between tables.

On the other hand, NoSQL databases provide greater flexibility by permitting horizontal scaling and storing data in a variety of forms, such as texts, graphs, key-value pairs, or column families. This makes them more appropriate for managing unstructured or semi-structured data. Consider it similar to the distinction between throwing things into various labelled boxes (NoSQL) and putting them neatly on shelves (relational).

It is frequently observed that rigorous transactional consistency may be sacrificed in order to gain flexibility and scalability in NoSQL databases. There are NewSQL systems for applicants looking for anything in between "Old Faithful" relational databases and NoSQL systems.

A recent effort that aims to combine the finest features of relational and NoSQL databases is called NewSQL. I describe how it is made to have the advantages of NoSQL without sacrificing the ACID characteristics of conventional relational databases. Because of its superior performance and scalability over traditional databases, as well as its consistency and dependability, NewSQL is a class of databases that combines the best features of both types of databases.

2.3 Selecting Appropriate Cloud Service Models (IaaS, PaaS, SaaS)

In response to shifts in the technology sector, the market for cloud service models has included a variety of new trends and technologies. These models are made to improve the viability, expansion, efficacy, and security of users' digital devices and content.

Therefore, getting to know them better and evaluating their capabilities appropriately are crucial when selecting the best cloud service model. In order to assist you with that, this blog post will analyse the three primary cloud computing models: Infrastructure as a Service (IaaS), Platform as a Service (PaaS), and Software as a Service (SaaS) (Mark R., 2023).

a. Infrastructure as a Service (IaaS)

Infrastructure as a Service (IaaS) is a popular cloud computing service idea that delivers virtualised resources via the Internet. By utilising the Infrastructure as a Service (IaaS) paradigm, cloud providers host the infrastructure components that are commonly present in an on-premise data centre. Among other things, these infrastructures may consist of servers, data centers, networking devices, and storage.

AWS, Cisco, IBM, Microsoft Azure, and VMware are just a few of the well-known hybrid cloud providers that enable customers to take use of IaaS in cloud computing.

Some essential features of IaaS use in cloud computing include:

- **On-demand resources-** In cloud computing, IaaS enables resource scale customization to satisfy needs. Because of these features, IaaS may be used by companies with different kinds of needs.

- **No physical hardware management-** IaaS companies ease customers' workloads by offering services like data center and server physical administration.

- **Location independence-** One of the key benefits of IaaS in cloud computing is its capacity to deliver services via the internet from any place. Thus, consumers are not restricted by their location when utilizing its services.

- **Flexibility and Control Over Infrastructure-** Although the infrastructure is provided by service providers, consumers can still alter features like the operating system, storage capacity, and installed apps.

- **Utility-Style Costing-** Customers can only pay for the services they use thanks to IaaS. Because of these benefits, IaaS is among the most well-liked options among hybrid cloud providers.

Typical Infrastructure as a Service (IaaS) use scenarios: In the industry, cloud deployment methods like IaaS are frequently employed for a number of reasons.

Here's a list of these use cases-

- **Storage and backup-** Data storage on cloud servers frequently makes advantage of its scalable methods. This enables IaaS users to store data securely and retrieve it whenever needed.

- **Testing and development-** To expedite the debugging process without having to purchase actual hardware, developers utilize Infrastructure as a Service (IaaS) to test their digital goods in a virtual environment.

- **High-Performance Computing:** Another reason IaaS is a popular cloud service paradigm is because of the resource-intensive computing environment. The program is utilised for important activities including data analysis.

b. Platform as a Service (PaaS)

PaaS is among the most popular types of cloud service models. The reason is the autonomy it grants app developers. App developers may create apps utilising PaaS without having to construct intricate infrastructures to support them. Alternatively, they can use PaaS-based infrastructures to develop and implement apps.

The capacity of PaaS to accelerate the entire development process is still the most significant advantage of cloud computing.

A few essential features of PaaS use in cloud computing:

- **Development framework-** Developers use the framework to create and modify mobile applications.

- **Integrated Development Environment-** Additionally, PaaS offers IDEs so that users may create apps on the server. The top cloud consulting firms also suggest PaaS to developers worldwide because of its testing capabilities.

- **Automated scalability-** PaaS is a well-liked cloud deployment methodology since it can automate the scalability element. It may change according on the website it is being viewed on.

- **Multi-tenant Architecture-** PaaS facilitates collaborations, enabling several users to access the application from various devices.

Common use cases of Platform as a Service (PaaS): Around the world, developers frequently take use of PaaS's advantages in cloud computing. Here are some typical use examples that may be used to learn how-

- **API Development and Management-** PaaS offers the ideal setting for making it easier to create, host, and administer APIs.

- **Application Development-** To speed up the app development process, PaaS provides a variety of pre-built backend infrastructure and development tools.

- **IoT infrastructure-** IoT infrastructure support is one of the main advantages of cloud service models like IaaS. Large numbers of IoT devices and their administration can be supported by PaaS.

c. Software as a Service (SaaS)

Software as a Service (SaaS) is a popular phrase for the process of developing software applications hosted by service providers. There is no need to install these apps on devices one at a time because users access them directly from web browsers.

Some critical characteristics of using SaaS in cloud computing:

- **Accessibility-** One of SaaS's most well-known features is its multi-device accessibility. It's a great tool for app developers because it can help a lot of people.

- **Centralized management-** The centralised management feature of SaaS makes it simpler to handle app data from a single location. It is also simpler to deploy fixes and upgrades for all app-using devices when administration is centralised.

- **Multi-tangency-** Software as a service (SaaS) for cloud computing may store user data separately for each user. This enables the provision of a more customised experience.

Common use cases of Software as a Service (SaaS): As previously said, SaaS is frequently chosen by app developers. This is one of the best cloud deployment techniques, and many companies have used it to improve their digital presence. Here are some observations.

- **E-mail and communication-** The ability to store emails and other digital communications is one of SaaS's main advantages in cloud computing. Data storage and interchange on virtual servers is made simpler by the cloud paradigm.

- **Customer Relationship Management (CRM)-** SaaS facilitates the storage of user information, preferences, and other attributes for an improved CRM system. Because of this feature, cloud servers are beneficial to even non-technical organisations.

- **Human Resources Management-** SaaS is being used by HR solutions to improve their hiring procedures. These businesses are using the advantages of SaaS in cloud computing to preserve pay scales, corporate data, employee data, and more.

2.4 Data Integration Strategies for Cloud Environments

Cloud integration data integration to be precise has become part of today's information technology foundations of any organization's operations. This process starts by adopting the hybrid integration methods where organizations can use on premise as well as cloud systems for integration. These operations are simplified by tools available in Integration Platform as a Service (iPaaS). With real-time updates between cloud and on-premise infrastructure, data silos are tackled effectively and access to consistent data is granted in all organizational processes. On the same note, integration through API allows numerous systems to come together by creating an ability to transfer data on real-time basis while API gateway hence offers major services comprising of security, monitoring, and versioning.

Another strategic area is data virtualization – the process of constructing integrated views on data without making actual copies. This approach minimizes feedback time and storage expenses yet provides completely direct access to data component. The same is true for ETL (Extract, Transform, Load) or ELT (Extract, Load, Transform) processes, which allow organisations to keep the specific data flows. ETL prepares data before loading in the target stage of processing while ELT loads raw data to cloud warehouses where it undergoes manipulation for analysis. The specific method to apply in data analysis depends on characteristics of the data set in question including size and complexity, and the need for processing. To support integration to a greater extent, there are specific tools, As examples of CN technologies with easy integration for automating and managing dynamic workloads, consider AWS Glue, Azure Data Factory, and Google Cloud Dataflow.

Another important characteristic that should be achieved in order to consider data integration successful is cloud data scalability and flexibility. The adoption of microservices architectures can enable the construction of organizational flexible systems towards responding to changes in business requirements. Multi-cloud and inter-cloud management initiatives are also crucial in order to run applications across different members of the multiple cloud platforms and in order to avoid dependency on a particular cloud supplier. Maintaining and establishing data quality, as well as adhering to data compliance and security requirements, which involve encryption and monitoring, are all made possible by data management regimes. AI and ML are two innovative technologies that help to advance integration since they decrease data mapping, solve errors, and optimize the workflow automatically. Prioritization of data quality, use of the incremental integration, as well as regular monitoring of the results allow organisations to unleash the advantages of cloud data integration and provide the market innovations that are sorely lacking in the current environment of intensifying competition in the data sector.

5 methods for data integration:

An essential component of building a successful data-driven business is data integration strategies. Companies may outperform their rivals by employing a range of integration strategies. A data integration plan considers the many data sources and kinds that your company has, the use cases and issues that arise during integration, and the platform and software that will be utilised to integrate the data. Organisations can use a variety of integration tactics based on their needs(Disney, 2023).

Data integration strategies assist in determining and implementing the best practices for information extraction, storage, and connection to corporate platforms and systems. Current data management and storage innovations, particularly cloud-based ones, are now taken into account by modern data integration strategies.

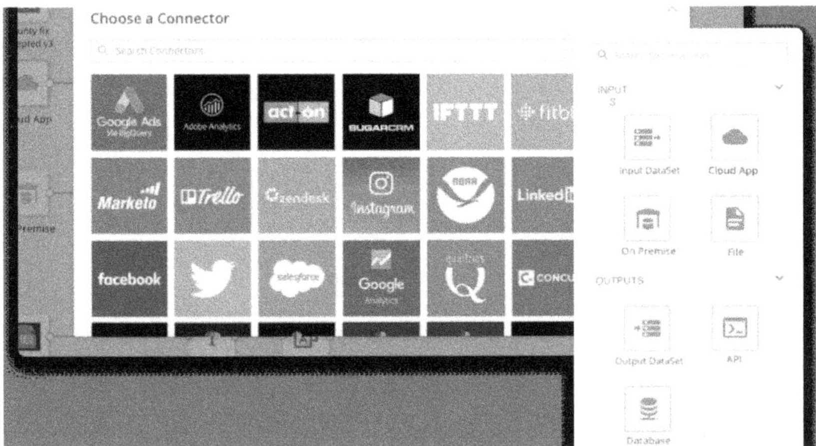

Source: - *(DOMO, 2023)*

The above figure "Choose a Connector" where one gets to select the integration type from Google Ads, Facebook, Salesforce among others. On the right is listed the input and output options: input dataset; cloud app; output dataset; API.

1. Integration based on applications:

Every step of the data integration process is managed by software applications in this approach. Data from various sources is identified, retrieved, cleaned, and combined by them. This automated approach makes it easy to move data across sources.

This approach is frequently referred to as enterprise application integration due to its popularity among companies that operate in hybrid cloud environments. These companies deal with a range of data sources, such as on-premises and cloud data sources.

Among the benefits of this approach are the following:

- **A simpler interchange of information**: An integration application may be used to transport information between departments and systems with ease.

- **Fewer resources are used**: The process is mostly automated, giving managers and analysts more time to work on other projects.

The following are some adverse consequences of this approach:

- **Inconsistent outcomes**: The service's methodology is not consistent and differs greatly throughout the businesses who provide it. You must learn more about the features of your organization's data in order to locate the integration app you require.

- **Restricted access**: The implementation of this approach necessitates technical expertise and the use of a data manager or analyst.

- **Challenging data management**: Multiple sources can occasionally cause issues since they compromise the data's integrity. To ensure the quality of the integrated data is double checked, someone should oversee the process.

2. Hands-on Data Integration:

This is known as manual data integration, when a supervisor performs or assigns each stage of the integration process, usually with code. This means manually gathering the data, connecting its many sources, and purifying it.

This method, although being a very time-consuming manual process, is effective. It may, however, soon become challenging for complex integrations. Everything is done by hand, costing money and time. This is known as manual data integration, when a supervisor performs or assigns each stage of the integration process, usually with code. This means manually gathering the data, connecting its many sources, and purifying it.

The advantages of this strategy include the following:

- **Lower system costs:** Manual data integration does not require costly hardware, software, or systems. Initially, entrepreneurs often start their businesses with a small budget. With a manual management method, a company may join the market without investing heavily in a powerful integration program.

- **Reliability**: In the case of a manual integration procedure, errors are less likely to continue since an employee is keeping an eye on every stage. Typically, workers will be able to identify errors as they happen.

3. Integration of middleware data:

One kind of computer program used to connect apps and transfer data across them is called middleware. When a business is merging stubborn historical systems with more contemporary ones, middleware may act as a bridge between different systems, which is very useful.

A few benefits of this approach are as follows:

- **Better streaming of data**: The middleware performs the integration automatically and reliably.

- **Streamlined system access**: Middleware makes it easier to link both ancient and new systems to a network.

The following are some of this strategy's shortcomings:

- **Experience required**: Middleware deployment task requires an experienced IT employee to handle the middleware.

- **Limited capabilities**: Some of them also lack compatibility with middleware.

Source: - *(Disney, 2023)*

The above figure represents a data visualization tool, of a data warehouse of some sort. They reveal several sources to and data sets in addition to metrics concerning the data and its current representation,

4. Data warehousing:

The data from many sources is stored in this approach in a data warehouse of a company or organization. Because of this, it is one of the frequently used types of data integration since it provides more freedom when manipulating data to businesses.

Among the identified advantages of this strategy the following can be mentioned:

- **Improved storage for analytics**: The construction of data and its storage in a data warehouse means that the managers and analysts are able to follow more complicated queries without worrying about overburdening the trigger databases.

Some of the demerits of this strategy are the following:

- **Increased storage costs**: The data must be stored somewhere and this particular type of data, namely data warehouse, will cost you some extra money, while at the same time, the majority of cloud-based storage options are not rather pricey. It also has to do with the size of data to be stored since more data is likely to cost more than a smaller amount of data.

- **Higher maintenance costs:** If the data warehouse is located on-site, technical staff must set up, manage, and run the server utilised for the data warehouse in order to coordinate the combination.

5. Integration of uniform access:

This keeps the data in its original location while retrieving it from several collections and displaying it in a single system.

This is the best approach for organisations that need to access many unique systems. This approach can yield insights without requiring the cost of data duplication or backups.

The following are some advantages of this approach:

- **Less storage capacity is needed**: To keep all of the organization's data, no special site has to be set up. This eliminates the need to pay for storage.

- **Easy access to data**: This approach works well with a range of data sources and platforms. Because a consistent access system shows the data while preserving its original sources, it makes data easily accessible. Storage of massive amounts of data does not reduce the capacity of the uniform access system.

- **A condensed view of the data**: The ultimate user sees the facts consistently, which facilitates comprehension.

The following are some of this strategy's shortcomings:

- **System strain**: In most cases, a large volume and frequency of data requests during this procedure may exceed some systems' capability.

- **Data integrity issues**: Utilising several sources and altering the data to present it consistently might jeopardize data integrity.

2.5 Ensuring Data Quality and Consistency

Reliability, correctness, consistency, and completeness of data are all considered aspects of data quality in business intelligence reporting systems. These dimensions are paramount in ensuring that the insights derived from a Business Intelligence reporting suite are sound and actionable,(Grow.com, 2024).

Industry leaders emphasize that the credibility of any BI reporting tool hinges on the quality of its data. A research by Gartner, for example, found that 40% of business activities fail to yield the desired results due to poor data quality.

Why Data Quality Matters

The completeness and accuracy of data determine its usefulness. Your company cannot genuinely make data-driven choices without accurate, pertinent, comprehensive, and consistent data, regardless of how many analytical models and data review procedures you use.

Understanding Data Quality

Data quality encompasses a variety of elements that affect the overall usefulness and reliability of a set of data. Having a solid grasp of the fundamentals of data quality can help you develop an action plan for setting and maintaining high standards.

What Is Data Quality?

A dataset's accuracy, completeness, consistency, and validity are all characterised by its data quality. Leaders must be confidence in the accuracy and dependability of their data in order to make data-driven choices. Inadequate data quality standards result in costly errors, inefficiencies, and business hazards.

Key Components of Data Quality

Teams must take into account data correctness, consistency, and integrity while assessing the quality of the data. It's also critical to consider the final application to make sure that the particular data points are pertinent to the precise issue you're attempting to resolve.

- **Data Accuracy:** Data is evaluated for how well it represents reality. Data errors like typos and outdated information reduce a dataset's accuracy. A high level of data accuracy requires few or no errors. By aiming for data accuracy, an organisation may make strategic data-based decisions.

- **Data Consistency:** Usable data is not just consistent but also correct. Consistent data is defined as coherent data across platforms and data storage. Teams working with data sometimes have difficulties with formatting and defining words to maintain uniformity, which lowers the information's usefulness. Teams may maintain data synchronization and raise the degree of confidence in that data by proactively and carefully creating interfaces and using tools like Where Scape 3D.

- **Data Integrity:** Therefore, data correctness characterizes the suitability of data as having been gathered throughout its collecting process in a thorough and consistent manner. Activities like relocating, merging, or remodeling data might inadvertently change datasets at any point during the data processing process. A compromised dataset is no longer complete or comprehensive and may be less useful for the purpose for which it was created.

Common Challenges in Maintaining Data Quality

a. Inconsistent Data Sources:

It often becomes a task to ensure data quality is maintained and one of the greatest issues that can be faced while using a Business Intelligence tool is that data compiled from different sources contain inconsistencies.

I enter data coming from various sources within an organization and in various formats, and in various quality and accuracy. For instance, when a multinational corporation consolidates data from regional offices into its main BI reporting platform, mismatched data formats and units will result in massive differences in the reports.

Solution: It is crucial to increase the likelihood of achieving data integration by following established protocols and to use a BI reporting tool that can maintain the data's data normalization across sources.

b. Data Duplication:

Redundancy is also a major problem in BI reporting tools. It manifests where a number of similar data entries are made a number of times with different entries but the same value. For example, in a sales database, the same customer may be entered more than once under slightly different names or phone numbers that complicate sales research.

Solution: Periodic assessments on the kind of data generated and the application of deduplication tools in the Business Intelligence reporting suite eliminates many of such duplications.

c. Outdated Information:

In this ever-changing business environment, information is dated as soon as it is gathered. A BI reporting tool depending of old data can cause errors that translate into unadvisable strategies. For instance, a retail company that feeds their Business Intelligence reporting suite with old customer preference data is likely to stock products that consumers have outgrown.

Solution: The other way of preventing horrible information in the Business Intelligence tool is by setting appropriate frequency for updating the data with new information and handling the data in real time.

d. Poor Data Quality at Source:

It is rather daunting that the phrase garbage in, garbage out applies well to BI reporting tools. One of the main problems that affect the BI process is data quality at its source. This issue is generally noticed in the traditional methods of entailment of data where errors are easily made.

Solution: The application of collection of data through Automation and applying stringent data validation at data capture stage can add value to Business Intelligence reporting suite.

e. Lack of Comprehensive Data Governance:

To maintain data quality in BI systems, a clear data governance architecture is actually necessary. The term "data governance" describes the guidelines, procedures, and best practices pertaining to the management and application of data. Lack of such can result in unstructured and ungoverned data usage, hence causing quality downgrades of business intelligence reporting tools (Atlan, 2023).

Solution: The major challenge involves creating a well-defined data governance program that covers all issues pertaining to data capture, utilization, analysis, and presentation in the Business Intelligence tool.

Best Practices for Ensuring Data Quality

a. Establishing a Strong Data Governance Framework:

The first line of defense in the handling of quality data in any BI tool is always data governance. This requires policy and procedural development and promulgations for data and information management. For instance, a financial institution may observe certain standards that are rigid when entering data in to the BI reporting tools in order to enhance accuracy. This governance framework should state responsibility for different processes and create an indicator for the correct approach to data(Rene Abraham, Jan vom Brocke, 2019).

b. Implementing Automated Data Validation and Cleansing:

A challenging element involved in manual operation is that data can easily be manipulated or even altered by human beings. It is possible to arrange data validation and data cleansing of inputs into your Business Intelligence reporting suite automatically, which drastically minimizes the occurrence of errors. BI reporting tools in automation are capable of not only pointing out errors, such as data inconsistency of the customers, but also rectify them on their own. You can use SQL Server Data Quality Services or Oracle Data Quality for this requirement that can be included in your Business Intelligence reporting suite of tools.

For more insights on efficient data management and the use of advanced tools, read our detailed blog Clean, Combine, and Conquer: This is all about enabling anyone and everyone to easily manage data through the help of Grow BI.

c. Regular Data Audits:

Data audits are therefore very important for ensuring that data in Business Intelligence tools are clean. These audits entail assessing and confirming compliance of the information used with expected standard of quality. For instance, a retail company may correct and verify its customer data kept in BI reporting tools for four times a year due to its nature of business.

d. Ensuring Data Completeness:

This makes it very easy for those who use business intelligence reporting to draw the wrong conclusions from incomplete data. Again, it is recommended that all the relevant fields in your Business Intelligence tool should be filled with all the appropriate information for best practice. For example, when all customer contacts are recorded and reported in the Business Intelligence reporting suite, better customer behavior assessment is made possible.

e. Emphasizing Data Accuracy at the Source:

The problem with BI reporting tools is that the quality of data attained in the BI reporting tools depends on the quality of data at the sources. That is why it is important that the data that is feeding your Business Intelligence reporting suite is clean. This could include educating employees on the entry of the right details or getting details in a way that minimizes the role of human actors.

f. Utilizing Advanced Technologies:

Implementing late-generation technologies like artificial intelligence (AI) and machine learning can even help business intelligence reporting solutions. These tools can provide a thoughtful grasp of aspects of data quality that an inspector might not even notice, as well as detect and even prevent data irregularities.

g. Fostering a Culture of Data Quality Awareness:

It is even more persuasive to make sure that all stakeholders in the company understand the importance of the high calibre of data that is received in the business intelligence reporting suite. This will be useful to ensure that BI tool staff enhance their knowledge and expectation concerning their contribution towards the improvement of data quality.

Advanced Techniques and Technologies in Data Quality Management

Advanced technologies such as Artificial Intelligence (AI) as well as machine learning are today transforming data quality in BI reporting tools. Data governance also has a major function in decisions and actions of a company. A data scientist, in an interview, explained to the writer, different AI algorithms that could predict and correct data anomalies in an ideal BI reporting suite(Praful Bharadiya, 2023).

Above is the story of Power blanket, a Utah based freeform heating products company that was struggling with its finance problem in the year of 2013 majorly because of using the old method of spreadsheet. Applying real time reporting at Grow changed a process of work with financial documents, decreasing time putting on it. When implemented, grow dashboards made it possible for them to increase employee productivity and satisfaction since they made available key information on their performance across their plant. This democratization of data and work streamlined reporting this led to Power blanket getting recognized in manufacture excellence all these show Grow's contribution to power blanket's efficiency as well as productivity.

Measuring and Monitoring Data Quality

Accuracy, completeness, and timeliness of data are the metrics used to evaluate data quality. Each team member must know their role in maintaining the security of the business intelligence tool, and there must be a culture that values high-quality data.

Implementing a Data Quality Framework in Your Organization

Building a data quality framework for your BI tool requires an in-depth understanding of your data requirements as well as the application of stringent data governance standards. You must educate your employees on the value of high-quality data in business intelligence reporting tools if you want to see results in the long run.

2.6 Security Considerations for Cloud BI

What is Cloud Security?

The phrase "cloud security" refers to the set of measures put in place to protect data, applications, and services that are hosted on the cloud. Compared to conventional on-premises systems, these environments—which include public, private, hybrid, and multi-cloud architectures—present additional security issues because they involve shared resources and dynamic infrastructure.

Cloud security encompasses a broad variety of defensive approaches, such as protection of networks, identification and access management (IAM), encryption of data, and detection of threats. Protecting sensitive data, preventing unwanted access, making ensuring regulations are followed, and shielding workloads from any breaches or assaults are its objectives(radware, 2022).

New cloud BI solutions have gained popularity over recent years because of their flexibility, affordability and availability. Yet, it also brings numerous security problems on the table which organizations need to solve in order to protect their data while keeping the trust in their analytics platforms. Such issues range from data privacy, regulatory requirements, privilege mechanisms, and many others.

Data encryption and storage Cloud BI has one of the biggest concerns regarding security of data that is stored and in transit. Any data which is backed up in the cloud should be encrypted using secure algorithms like Advanced Encryption Standard – 256. Furthermore, organizations have to make sure that data that is transmitted between the organization and the different entities is protected by using protocols such as the TLS (Transport Layer Security). Encryption keys have to be kept secure, and many providers can provide Key Management Services (KMS) to allow businesses to dictate the usage of these keys.

Directory Access Management Since Cloud BI systems involves personal data of an organization, the significance of directory access management cannot be underrated. It is recommended that IAM should be enriched in organizations by the establishment of policies that will put a check on any unauthorized access. This includes RBAC to restrict user from gaining access to data as well as tools, which they have no business interacting with. Using of the multiple-factor authentication helps to minimize the possibility of unauthorized access.

Notable ones are: Compliance and Regulatory Conformity It is common for many industries to be bound by often-strict requirements regarding how data must be processed and used; these governing regulatory frameworks include GDPR, HIPAA, and CCPA. Cloud BI providers must recognize that organizations need tools and certifications to remain legal. It is important for organizations to check their chosen provider meets these regulations and have the proofs that they do.

Data Ownership Data governance and ownership organizations bear the question of who owns the data that's collected as well as processed in a cloud BI platform. As for the subject of ownership and usage of the data, it is recommended that this issue to be clarified in service agreements. Moreover, using data governance frameworks extends the idea of managing data and ensures that data is well aligned to an organizations' policy.

Vulnerability Management and Incident Response Cloud environments are not isolation from having vulnerabilities. The BI in the cloud should be tested for security vulnerabilities through conduct of security audits, penetration testing, and patch management should be conducted as often as possible. Organizations must clearly provide a detailed course of action that must be taken in the wake of an attack. Organizations should also prepare internal incident response that are specific to their business intelligence processes.

Data Backup and Disaster Recovery, It is also an important area of securities that relates to BI, which confirms the availability of data that have to be recovered if necessary. The providers should provide a backup of demands related to automation and clear-cut disaster recovery plan. Companies need to identify contingency objectives, such as RTOs and RPOs, coming from the cloud BI vendor according to the business continuity plan.

In a cloud BI model, security concerns are shared between the two parties; the provider and the customer. It is common for the formers to take care of the cloud infrastructure, whereas the latter's are to ensure security of their data, configurations and users. Indeed, when it comes to using technologies on the web it is important to know this model as it will help in eliminating possible security vices.

Security of Cloud BI Solutions Reliability and Security Practices Reliability and security practices characterised a cloud BI vendor solution's fundamental security model. The organizational factors that can be used as indicators by organizations while choosing a specific vendor are the records of the vendor, certifications, whether they have received ISO 27001 or SOC 2, and third-party audit reports. Security measures and policy statements should always be transparent so that the clients have a reliable channel to approach when they want some clarifications about the security measures to be undertaken on their accounts.

User Training and Awareness: being a critical problem human error is definitely one of the most significant contributors to security breaches. Reduction of the risks can be done frequently through training and awareness programs for members using the cloud BI system. People within an organization should be taken through an awareness program on phishing, the right passwords and proper security policies to follow.

Although employing Cloud BI has numerous benefits, there are security dangers that businesses should be aware of. Data vulnerability or hacking of the cloud infrastructure where this data is housed is one of the main issues.

Throughout the years, several cloud-based systems have been targeted, some of which were quite visible, highlighting these dangers. One likely example is the 2019 data breach at Capital One, when the company's AWS environment's improperly configured web application firewall allowed the theft of over 100 million customers' personal information.

These incidents need to encourage businesses to participate in the deployment of security measures in addition to those implemented by cloud service providers.

Privacy Concerns in Cloud BI

In addition to security risks, privacy concerns should be taken into account while deploying Cloud BI. Privacy regulations should be in place to safeguard the data being used and prevent access by third parties.

Among these is the potential for cloud service providers and other third parties to violate data confidentiality. Although cloud service companies have safeguards in place to protect their customers' data, insiders or workers who are legally permitted to operate in the company's environment pose a serious risk.

Additionally, a number of legal and regulatory concerns, such data protection laws, continue to worry businesses considering Cloud B adoption.

To a certain degree, the laws of various nations and areas have been formed differently, and as a result, the regulations governing the processing of personal data also vary. In addition to the potential for very severe fines, noncompliance with these standards may harm a company's reputation.

Figure 2.1: A cloud computing network with data transfer and storage icons."

Source: - *(fintrak, 2023)*

Compliance and Regulatory Considerations for Cloud BI

When deploying Cloud BI, Azure, Amazon Web Services, and Google Cloud are some of the clouds that must be given priority. Privacy considerations must also be considered in accordance with the rules and frameworks that apply to the particular organisation.

Limits on data collection, storage, processing, and sharing have been imposed by legislation such as the EU's General Data Protection Regulation (GDPR). Businesses with operations in EU member states or that deal with personal data of EU residents are required to adhere to these regulations.

Similarly, the United States and other nations have legislative tools like HIPAA to protect health information. According to these regulations, businesses handling protected data in the cloud environment must abide by specific guidelines.

Businesses interested in deploying Cloud BI must first thoroughly assess the degree of compliance required based on the business vertical and geographic location before transferring their vital data to the cloud.

Best Practices for Securing Cloud BI

In order to mitigate security concerns during Cloud BI implementation, the organisation should follow best practices for system security that the BI solution will use:

a. Implement strong authentication mechanisms:

Strong authentication methods must thus be used in order to increase the security of such data in the context of cloud BI. There are methods for using many-factor identification validation, such as using tokens or fingerprints, to safeguard the privacy of the data.

Because biometric authentication relies on unique characteristics like a person's face or fingerprints, it is completely proof because only the biometrics' true owner can perform it. Token-based systems also generate one-time passwords or tokens that users must possess in order to access BI tools and analytics.

b. Regularly update software:

Updating specific components on a regular basis ensures that the system and its users are protected from various dangers. This process becomes even more important when data is handled utilising several platforms and sources and when company information is stored in business intelligence systems that are hosted in the cloud.

Organisations can protect themselves from the ever-growing threat of cybercriminals and other unauthorized individuals trying to infiltrate their cloud space and obtain sensitive data that belongs to them by updating their security patches. Due to the possibility of using several data sources, cloud BI platforms must improve the security of all linked systems.

c. Encrypt data in transit and at rest:

Cloud services are quickly becoming the go-to option for hosting BI tools and data. Therefore, it is crucial that players cannot access the data in order to prevent private information from being disclosed to third parties. Ensuring the security of both moving and stationary data is crucial. Encrypting data ensures its security during transmission and storage on cloud servers.

This makes it very hard for the individuals who might otherwise penetrate the network to be able to understand the intercepted data.

d. Conduct regular security audits:

Cloud businesses have to focus on security, starting with offering routine security assessments of their infrastructure. This include using tools and services in evaluating the exposure of their cloud data. Therefore, in penetration testing vulnerabilities in the system is exposed before it reaches the hands of the attacker.

The centralization of the cloud environment helps to monitor, and one way of achieving it is by having a dashboard view of the environment. This also makes it very easy to provide a relevant response in case there is any threat to the security of the organization. These audits are important to keep updating businesses since it reduces risk posed by cyber threats and boosts overall security.

It also assists in keeping track of compliance to an industry's regulations and standards so as to protect and secure privacies from loss or theft. Ensuring sound security practices are established and employed is a sign of appreciation to customers as well as being an effort to ensure there is credibility in the organization.

Securing Cloud Business Intelligence

Cloud business intelligence security is one of the most important issues of the current management of data. Since organizations invest heavily in the use of cloud-based tools and services for managing and analyzing data, then security of data incurs high importance. One of the main areas of interest is access to cloud data with a focus on generating efficient mechanisms of authentication and encryption.

- **Encryption and Data Protection in Cloud BI:**

  In Cloud BI architecture, secure encryption is vital as it helps in protecting the knowledge in the BI system. It comprises the process of converting readable data in the normal format into an encrypted format which cannot be comprehensible in the absence of authentication key.

  Cloud BI incorporates the following basic encryption methodologies namely the symmetric, asymmetric encryption and hashing algorithms.

  In symmetric encryption a same key is used for both encryption and decryption of data or information. This method is efficient though the key needs to be shared securely between the players involved.

Asymmetric encryption has two unique key – the public key which encrypts data and the private key that decrypts the data. This approach is also more secure because while the private key remains out of the public view everyone who has the public key can encrypt information.

There is a process of hashing, which creates a digital or computer representation of any data in the form of a hash value distinctive from other data. These hashes can only be used to commend files for integrity or to compare files without having to know the content of the files.

An organization using good encryption tactics means that even though an attacker might intrude into the cloud systems or capture data transmission, the data will remain encrypted such that an attacker will not understand or access the information.

- **Access Control and Authentication in Cloud BI:**

  This is important in Cloud BI systems as access control methods ensure only right people get access to certain critical information.

  This means that organizations should use proper methods of authentication for instance MFA or SSO. MFA adds another layer of authentication, over and above a username and password, as the user is required to give multiple forms of identification before using the resources.

  SSO enables a user to log into any given number of systems or applications without requiring him/her to enter several forms of credential information. The reasons with this are because weak or repeated passwords are usually some of the most common security threats.

  Also, organizations should use RBAC which restricts access for each user depending on his/ her position within the organization. RBAC makes it possible that any person is allowed to access only the information and the functions that they will require in performing their duties.

- **Monitoring and Auditing in Cloud BI:**

  The paper shows that the monitoring and auditing are crucial elements for the Cloud BI security plan. These practices assist an organization to identify likely security breaches, risks and compliance gaps.

  The organisation can often keep an eye on what's happening in the cloud environment thanks to cloud service providers. Dynamic control of system and network logs, user activity, and other factors is provided by these technologies. By using these logs in conjunction with system events, organisations may determine whether or not there is any unusual behaviour that could indicate efforts at intrusion or unauthorized persons attempting to access the system. It makes it possible to swiftly start reaction actions that can stop more impacts from happening.

  Auditing consists of carrying out routine inspection of an organization's cloud environment with the objective of its conformity with certain laid down security measures or a set of standards. It helps to review all controls and finalize that all are functioning enough and are implemented.

2.7 Performance Optimization Techniques

Cloud BI solutions have become an essential tool that companies that seek to make sense of their data through Business Intelligence need. But this comes with a challenge of how the current expansive and complicated data can still support high performance. Effective Cloud BI solutions imply that data processing, the analyses, and visualizations needed for management decision-making are done effectively to cover both ongoing and managerial decision-making requirements. Applying the trends for improving the data processing, computational capabilities, and system expansion, organizations can offer the BI experience that users will expect.

Figure 2.2: Performance Optimization

Source: - *(Shah, 2017)*

a. Optimizing Data Modeling and Storage:

Data modeling forms the single most important foundation of any superior BI architecture. A good structure will help in querying and take less time to process than a model with poor data structures. Working with data, simple star schema design well translates into analytical fact and dimension tables for managing intricate workloads. Further, the data are beforehand aggregated if and when similar metrics are conducted frequently in order to minimize the amount of time computation takes when analyzing data. Data storage optimization is also equally important. Group: Storage Optimization Choosing columnar storage formats such as Parquet or ORC improves the read speed of analytical queries because both formats were created for sequential reading. If appropriately used, the compression reduces the cost of storage and enhances I/O operations through footprint reduction. Division of large files by time or geographic location makes data access faster when it is needed for time-series or regional analysis,(Tech target, 2024).

b. Streamlining ETL/ELT Pipelines:

Usually, the data gathered undergoes the Extract, Transform, Load (ETL) or sometimes Extract, Load, Transform (ELT) processes. Improved efficiency and cost are some of the dramatic changes achievable through optimizing these value delivery pipelines. An on-going data ingestion process is achieved through incremental processing where only updated records are processed resulting in faster data ingestion. Transformation pushdown—realizing transformations on the data warehouse—minimizes internal data movement, taking advantage of the computational capabilities of contemporary cloud-born warehouses. This shows that parallel processing, which is designed to accomplish large data tasks by breaking them down into many simpler, simultaneous data alteration jobs, guarantees scale and the capability of delivering the data on time without interruption due to excessive workload. All such optimizations collectively act to minimize latency and optimize the overall throughput of data pipeline.

c. Query Performance Tuning and Dynamic Resource Management:

Query performance exactly impacts the user experience in context to BI systems. Improving the execution of SQL queries can be made easier by avoiding extraneous joins, using filters strategically and minimizing on the use of nested subqueries. This procedure, using different indexing techniques involves creating clustered and non-clustered indexes which highly increase the efficiency of the query retrievals. Thus, materialized views, which contain results of prior queries, are simply inestimable for repeated use, when one does not need real-time computation. Cloud platforms have dynamic resource management attributes such as autoscaling which enables a provider to scale up or down the compute and storage resources it provides depending on the load encountered. In addition, using cache mechanisms, including local and distributed ones, can enhance query due to the lowered database access frequency for frequently used data.

d. Leveraging Cloud-Native Tools and Features:

Snowflake, Google Big Query and Amazon Redshift type cloud native data warehouse is optimized for performance and come with tuning options to configure on top of them. Some of these platforms have nice- to-have capabilities such as autoscaling, where a system adjusts to concurrency in a bid to exhibit a stable performance regardless of traffic surge. Through workload management, administrators can directly deal with important queries, and allocate their resources to where the most need is. Most of these also give the query optimizer feedback and suggestions derived from historical use of the queries. Through these tools, organizations can realize large improvements in performance without much of the hands-on support, enabling BI groups to concentrate on providing information instead of managing infrastructure.

e. Monitoring and Continuous Optimization:

Optimization of performance is therefore not a once in a while kind of activity but a continuous one. This is because regular audits of the system should reveal tendencies that show areas of weakness or points of congestion. AWS and Azure monitoring services give fine-grained observations of system health, resource usage of the solutions, and query effectiveness. Using query performance analysis tool, one can diagnose slow queries while log analysis diagnoses general patterns and trends that may affect the performance. The workload patterns are best to be predicted and need to be integrated into the machine learning models to facilitate the preemptive resource assignment and identification of any abnormities. This continuous feedback loop makes the BI solution to be dynamic and capable of meeting the increasing volume of data and the increasing number of users.

f. Balancing Security, Governance, and Scalability:

However, the performance optimization requires strong security and compliance as the primary foundations. It is possible to provide row-level and column-level security measures to secure a database but with slightly reduced query performance. The overnight consistency and validity checks are also carried on an automated manner which only allows proper data into the system. Scalability has to be an important factor, especially as these solutions are often realized in cloud environments. Using machine learning, Spree scaling can learn from past experience and proactively load up additional resources to prevent subpar performance during crucial moments. Bi performance, security and scalability are maintained at optimal level to meet the business needs while at the same time upholding and being compliant to set rules and regulations.

2.8 Leveraging the Analytics Pane: Advanced Tools in Power BI Desktop

Power BI Desktop's Analytics pane allows you to highlight key trends or insights, add dynamic reference lines to visualisations, and more. Within Power BI Desktop's Visualisations section, you'll find the Analytics icon and window.

Figure 2.3: Use the Analytics pane in Power BI Desktop

Source: - *(Davidiseminger et al., 2023)*

Search within the Analytics pane

Within the Visualisations window, namely the Analytics pane, users have the ability to conduct searches. Clicking the Analytics symbol brings up the search bar.

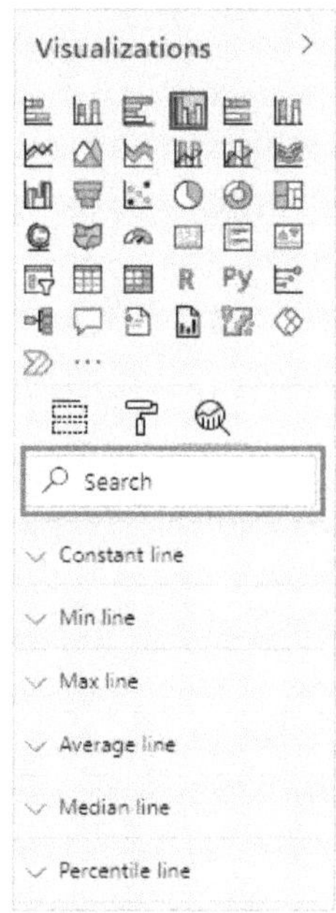

Figure 2.4: Use the Analytics pane in Power BI Desktop

Source: - *(Davidiseminger et al., 2023)*

Use the Analytics pane

In the Analytics pane, you can create dynamic reference lines like these:

- X-Axis constant line

- Y-Axis constant line

- Min line

- Max line

- Average line

- Median line

- Percentile line

- Symmetry shading

How to incorporate the Analytics pane and dynamic reference lines into your visualisations is demonstrated in the sections that follow.

Here are the steps to see the visual's available dynamic reference lines:

1. In the Visualizations section, find the Analytics icon after selecting or creating a visual.

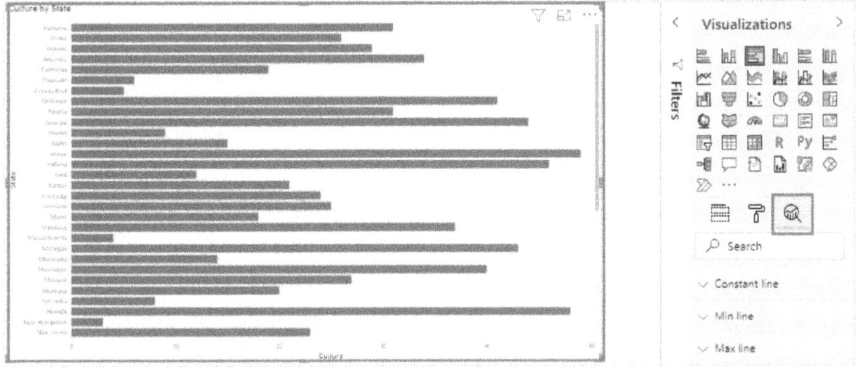

Figure 2.5: Steps Use the Analytics pane in Power BI Desktop

Source: - *(Davidiseminger et al., 2023)*

2. To access its choices, choose the line type you wish to create. The chosen Average line is displayed in this example.

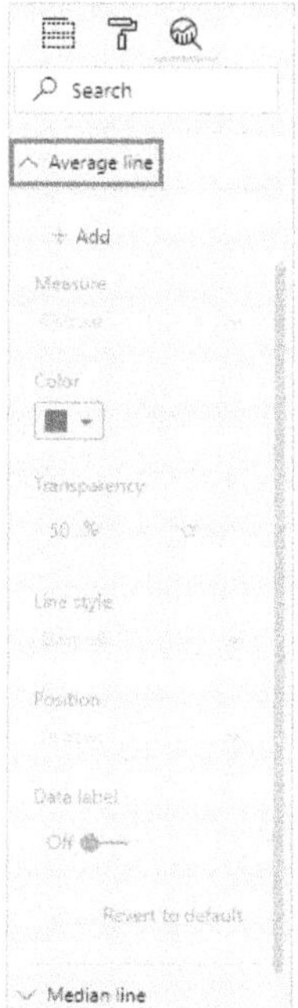

Figure 2.6: Steps Use the Analytics pane in Power BI Desktop

Source: - *(Davidiseminger et al., 2023)*

3. Choose + Add to add a new line between two lines. And after that, you can give the line its name. Type your name into the text field by double-clicking it.

    A plethora of choices are now available to users for your queue. In addition to positioning it relative to the data items in the visual, you can also choose its colour and transparency percentage. The Data label is an additional option that you can select. By default, data items from the visual are filled into the Measure dropdown selection, so you can choose it to use as the basis for your line. In this case, we'll show you how to calculate the average of culture using the Culture metric. Along the way, you'll find ways to personalise a couple of the other settings.

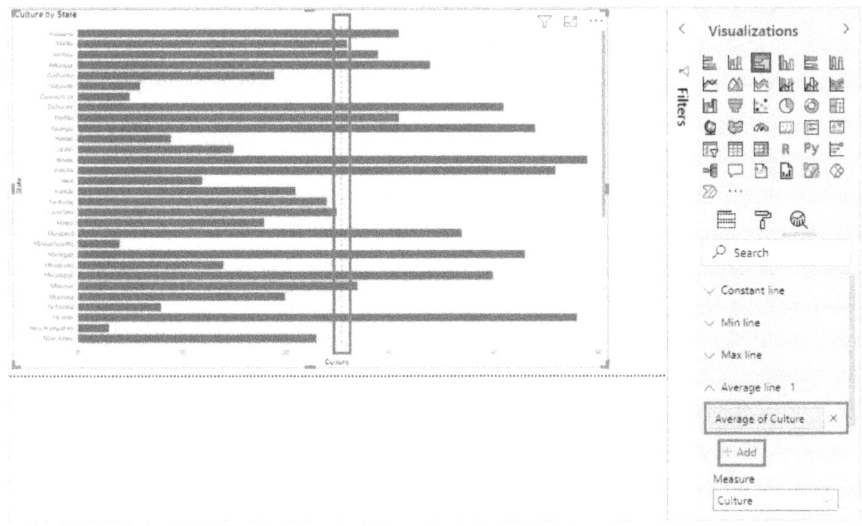

Figure 2.7: Steps Use the Analytics pane in Power BI Desktop

Source: - *(Davidiseminger et al., 2023)*

4. Turn on Data label if you'd like a data label to show. There are a lot more choices for the data label when you do this.

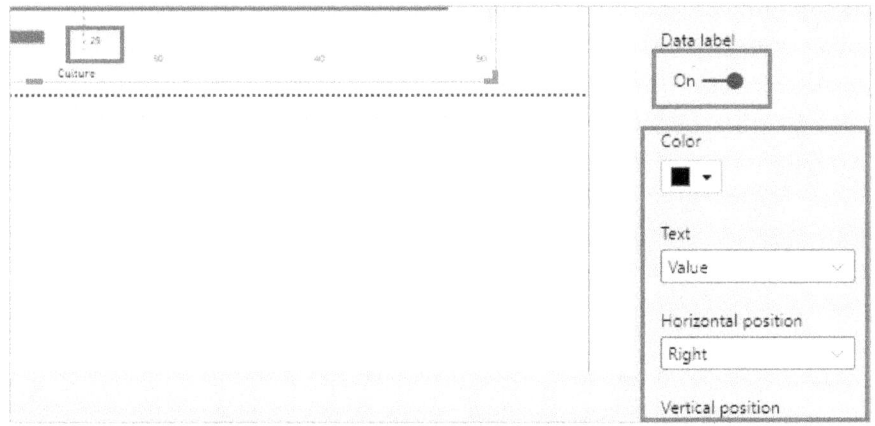

Figure 2.8: Steps Use the Analytics pane in Power BI Desktop
Source: - *(Davidiseminger et al., 2023)*

5. Look at the Analytics window; there should be a number next to the Average line item. That will inform you of the type and number of dynamic lines present in your image. Incorporating a Max line for Affordability into this visual also applies a Max line dynamic reference line, as seen in the Analytics pane.

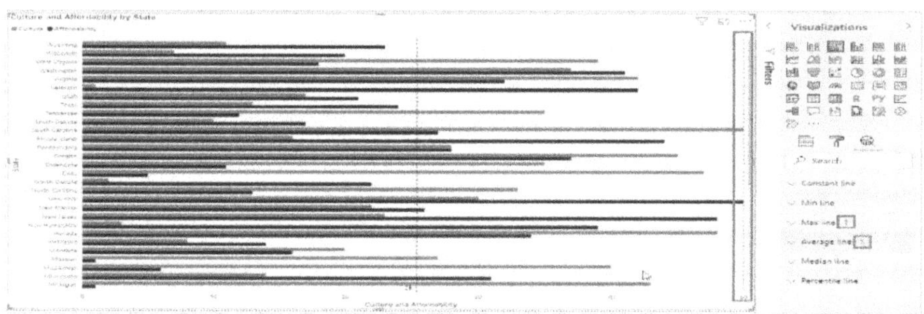

Figure 2.9: Actions Taken by the Power BI Desktop Analytics Pane

Source: - *(Davidiseminger et al., 2023)*

When you choose the Analytics pane, you'll receive this alert if the visual you've chosen (a Map visual in this example) can't have dynamic reference lines attached to it.

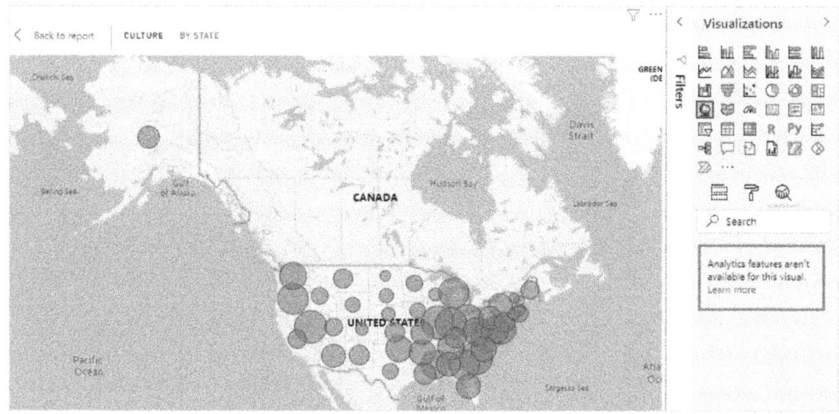

Figure 2.10: Procedures for Using Power BI Desktop's Analytics Pane

Source: - *(Davidiseminger et al., 2023)*

In the Analytics pane, users can draw attention to numerous noteworthy facts by generating dynamic reference lines.

2.9 Chapter Summary

This chapter examines how to create, deploy, sustain and protect agile, efficient and secure cloud BI systems. Flexibility of data structures is stressed by horizontal and vertical scaling, partitioning, and sharing, in addition to using relational, NoSQL, and NewSQL databases. IaaS, PaaS and SaaS models of cloud services are described separately with emphasis on what they offer and require. Data integration issues are described in the chapter with reference to hybrid integration approaches, middleware, and data virtualization, as well as the quality and integrity of data used in BI. It covers issues of encryption, compliance, and access as well as solutions to risks like data breach. Other options for improving performance that cover query tuning, resource management, and use of CN-Specific Tools are also described to fulfil more dynamic, successful, and scalable BI solutions.

Multiple Choice Questions (MCQs)

1. **What is a primary goal when designing scalable data architectures for Cloud BI solutions?**

    a. Reducing data redundancy

    b. Maximizing data silos

    c. Avoiding data backup processes

    d. Using only on-premise hardware

2. **Which cloud service model provides the highest level of control over infrastructure?**

    a. SaaS

    b. PaaS

    c. IaaS

    d. FaaS

3. **What is the key advantage of using PaaS in Cloud BI solutions?**

    a. Complete control over hardware

    b. Simplified application development

    c. Enhanced end-user interfaces

    d. Full hardware management by the client

4. **Which of the following is NOT a common data integration strategy for cloud environments?**

    a. ETL (Extract, Transform, Load)

    b. ELT (Extract, Load, Transform)

    c. Data virtualization

    d. Data segregation

5. **What is a critical aspect of ensuring data quality and consistency in Cloud BI?**

    a. Ignoring data validation

    b. Relying solely on manual processes

    c. Implementing data profiling and cleansing

    d. Avoiding metadata management

6. **Which security measure is most critical for protecting sensitive data in Cloud BI solutions?**

    a. Disabling encryption

    b. Enforcing multi-factor authentication (MFA)

    c. Using shared passwords across teams

    d. Ignoring compliance standards

7. **What technique can help optimize performance in Cloud BI environments?**

    a. Overloading servers with concurrent queries

    b. Implementing caching mechanisms

    c. Disabling autoscaling features

    d. Using outdated hardware configurations

8. **What is one of the key considerations when selecting a cloud service model for a BI solution?**

    a. The proximity of the cloud data center to the end user

    b. The number of employees in the organization

    c. The level of customization and control required

    d. The type of mobile devices used by employees

9. **What is a primary reason for implementing data integration strategies in Cloud BI solutions?**

    a. To increase data fragmentation

    b. To consolidate and unify data from multiple sources

    c. To reduce the use of ETL tools

    d. To minimize cloud storage costs

10. **What is typically included in the summary of a chapter on architecting Cloud BI solutions?**

    a. A detailed list of cloud providers

    b. Key takeaways and highlights of the discussed topics

    c. A guide to building physical servers

    d. Instructions for creating non-scalable architectures

Answer

1	2	3	4	5	6	7	8	9	10
c	c	b	d	c	d	b	c	b	b

Chapter 03

IMPLEMENTING CLOUD BI IN HEALTHCARE

3.1 Chapter Overview

This chapter looks at how organizations in the healthcare industry are utilizing cloud-based business intelligence to deliver information that enhances patient and organizational performance. They explain how cloud BI can be used to improve the delivery of patient services by leveraging analytical technologies, meet important healthcare regulatory standards including HIPAA, and connect EHR for consistency across data platforms. Further, the chapter builds the promise of PA for disease regulation, identifies issues and their resolutions for cloud BI adoption in healthcare, and provides a comprehensive real-world hospital network success story. The intended audience of this paper will be able to comprehend the different ways cloud BI can foster enhanced innovation within the healthcare sector by the end of this study.

3.2 Enhancing Patient Care with Data Analytics

One of the most promising areas to investigate with data analytics in this era of data-driven decision-making is healthcare. With healthcare systems worldwide undergoing constant change, data leveraging to boost patient outcomes, optimise operations, and improve healthcare delivery as a whole is gaining importance. By altering how healthcare providers collect, analyze, and utilize data, healthcare data analytics can radically improve patient care. Learn about the many kinds of data analytics, how they help healthcare, and how they will change the way patients get treatment in the future in this blog post (Dash et al., 2019)store, and analyze big data with an aim to improve the services they provide. In the healthcare industry, various sources for big data include hospital records, medical records of patients, results of medical examinations, and devices that are a part of internet of things. Biomedical research also generates a significant portion of big data relevant to public healthcare. This data requires proper management and analysis in order to derive meaningful information. Otherwise, seeking solution by analyzing big data quickly becomes comparable to finding a needle in the haystack. There are various challenges associated with each step of handling big data which can only be surpassed by using high-end computing solutions for big data analysis. That is why, to provide relevant solutions for improving public health, healthcare providers are required to be fully equipped with appropriate infrastructure to systematically generate and analyze big data. An efficient management, analysis, and interpretation of big data can change the game by opening new avenues for modern healthcare. That is exactly why various industries, including the healthcare industry, are taking vigorous steps to convert this potential into better services and financial advantages. With a strong integration of biomedical and healthcare data, modern healthcare organizations can possibly revolutionize the medical therapies and personalized medicine.","author":[{"dropping-particle":"","family":"Dash","given":"Sabyasachi","non-dropping-particle":"","parse-names":false,"suffix":""},{"dropping-particle":"","family":"Shakyawar","given":"Sushil   Kumar","non-dropping-particle":"","parse-names":false,"suffix":""},{"dropping-particle":"","family":"Sharma","given":"Mohit","non-dropping-particle":"","parse-names":false,"suffix":""},{"dropping-particle":"","family":"Kaushik","given":"Sandeep","non-dropping-particle":"","parse-names":false,"suffix":""}],"container-title":"Journal   of   Big   Data","id":"ITEM-1","issued":{"date-parts":[["2019"]]},"title":"Big data in healthcare: management, analysis and future prospects","type":"article-journal"},"uris":["http://www.mendeley.com/documents/?uuid=1b89cd20-7f92-4465-a29b-9e5d31c6fc05"]}],"mendeley":{"formattedCitation":"(Dash et al., 2019.

What Is Healthcare Data Analytics?

By methodically examining health data, healthcare data analytics aims to improve patient care, simplify operational processes, and supply better information for strategic decision-making. By sifting through mountains of administrative, clinical, and financial data in search of trends, healthcare organizations can enhance provider and patient outcomes through evidence-based decision-making (Khan et al., 2022).

When it comes to healthcare, data analytics is more than just gathering data; it's about deciphering complicated databases to find insights that can improve efficiency, effectiveness, and personalization of care. Healthcare data analytics improves the quality of care and allows for better allocation of resources, which in turn reduces hospital readmissions and increases preventive care measures.

Types of Healthcare Data Analytics

Each of the four primary categories of healthcare data analytics contributes to the overall system in its unique way. To better comprehend how data might guide and enhance patient care, these kinds offer a framework.

- **Descriptive Analytics.** Descriptive analytics involves the use of historical data to make past event explanations. The function of this could be, for example, analyzing patients' medical records, their treatment record and the outcomes obtained from their specific treatments in the healthcare sector'. For instance, descriptive analytics may be used in a hospital to determine trending or perennial admitting diagnosis in the preceding year or to monitor infections in varying units.

- **Diagnostic Analytics.** Diagnostic analytics takes it a notch higher by examining data and finding why specific occurrences happened.

Aiming to provide a response to the query "Why did this happen?" In the medical field, for example, diagnostic analytics can help determine why certain patients have a higher risk of complications after surgery or why certain demographics are more susceptible to certain diseases. If the health care industry wants to know what else is causing certain health outcomes, they can employ this kind of data analysis.

- **Predictive Analytics.** Data, statistical methods, and machine learning are all part of predictive analytics, which aims to make predictions about the future. Medical facilities can therefore anticipate patient needs, illness risks, and general health hazards thanks to the massive amounts of data kept. For instance, risk assessments can be employed to predict those learners most at risk of coming up with chronic diseases using their health status and lifestyle.

- **Prescriptive Analytics.** While prescriptive analytics can tell organizations, what is likely to happen in the future of a model, it prescribes actions that must be taken in return. In healthcare, prescriptive analytics could recommend the right treatment plan for a patient, taking into account the patient's history, the patient's current health and details from other treated similar patients. This kind of analysis assists hospitals, clinics or any other healthcare organizations to come up with right decisions on patient care patterns and resource utilization as well as the best approaches to adopt in clinical practice.

Benefits of Data Analytics in Healthcare

Data analytics has many uses in healthcare and can benefit both organisations and patients. Decisions made with data in hand have the potential to enhance clinical care experiences, operational efficiency, and patient outcomes in healthcare systems. Here are only a handful of the most significant benefits:

- **Improved Patient Outcomes:**

  The potential of healthcare data analytics to enhance patient outcomes is an important benefit. Health care providers might detect possible problems or dangers by examining patient data for patterns. Personalized treatment plans, earlier action, and improved diagnostics are all made possible by this. To illustrate the point, physicians can utilize predictive analytics to spot patients who are at a high risk for developing chronic conditions like diabetes or heart disease, and then take preventative steps before the problem gets worse. There is a way for healthcare providers to be more proactive and preventive with their patients' health with the use of a predictive healthcare model.

A patient's specific genetic makeup and medical history inform the best course of therapy, which is another benefit of prescriptive analytics. A new paradigm in healthcare with the potential to revolutionize treatment and health outcomes is precision medicine, which takes a patient-centered, individualized approach.

- **Enhanced Operational Efficiency:**

  By illuminating inefficient processes and providing avenues for process simplification, analytics of data can also assist healthcare organizations in enhancing their operations. For instance, healthcare facilities can enhance resource allocation, staffing management, and patient demand forecasting with the help of data analytics. Analyzing patient flow data allows healthcare organizations to optimize operating room schedules, improve bed management, and reduce wait times.

  The rising expense of healthcare can be mitigated by the use of data-driven insights to cut down on needless medical procedures and hospital readmissions. More efficient and less expensive treatment can be achieved when healthcare organizations concentrate on enhancing procedures and decreasing waste by discovering patterns in patient data.

- **Preventive Care:**

  Data analytics is essential to the development of preventive care, which is a primary emphasis in contemporary healthcare. Medical professionals can use data mining to find potential danger spots and foretell which individuals will get certain diseases. This means patients can avoid unnecessary wait times for treatment, which boosts their quality of life and reduces healthcare costs.

  As an example, healthcare organizations can utilize predictive analytics to pinpoint groups that are more likely to suffer from conditions like obesity, diabetes, or hypertension. Providers can use this data on the population's health to launch specialized programs of preventative treatment, such as health education, behavioral modifications, and screenings, with goal of lowering the prevalence of these diseases.

- **Improved Patient Engagement:**

  Data analytics is also having a major effect on patient involvement. Healthcare providers can encourage patient engagement in their care by making patient records available online through patient portals, smartphone applications, and other technological means. The result is enhanced two-way communication between patients and doctors, which in turn improves patients' compliance with their treatment programs and, eventually, their health.

  Healthcare organisations can also use data analytics to tailor patient communications to individual tastes, medical histories, and treatment plans. For instance, we may improve engagement and the patient experience by customising automated reminders for patients' medication refills, follow-up appointments, and preventive screenings.

A Health Data Analyst's Function

The main work of the healthcare data analysts is to take the raw form of the health data and turn it into insights. Their key roles involve generating, capturing, converting and analyzing vast amounts of data from different sources including EHR technologies, medical equipment and patient feedback, insurance claims. Through the interpretation of these difficult datasets, healthcare data analysts said the providers in making economically sound decisions which would translate to enhanced patient experience, increased effectiveness and, reduced expenses.

The first prerequisite of a professional healthcare data analyst is that he/she must have a good grounding in data science, statistics and in healthcare management. He co-ordinates with doctors, clinics, hospital authorities and professionals involved in medical informatics to create foresight, improve productivity and make statistics accountable and responsible at the end.

However, health care data analysts should also have a strong base of subject matter knowledge such as knowledge of health care institutions, policies, procedures and rules and regulations regarding patient care among other things. A key aspect of work of Data Governance Committees is to facilitate the use of data analytics tools and methods that are appropriate with regard to the objectives of an organization or needs of patients.

Because more healthcare organizations generate revenues from providing patient care and minimize costs from academic missions, a focus on data-driven, research-based decisions will only increase the need for skilled healthcare data analysts. These professionals will play special roles in assisting healthcare systems in understanding the challenges of big data and applying them to innovate and advance the delivery of health care. Healthcare data analysts are a crucial part of healthcare informatics since they seek to bring about the patient experience with the help of data analytics.

The Future of Patient Car with Healthcare Data Analytics

With data analytics, patient care has the potential to undergo a monumental transformation, and data is driving the future of healthcare. More and more creative uses of data analytics in healthcare are likely to emerge as technology develops further. These may include:

- **Artificial Intelligence and Machine Learning:** Algorithms driven by artificial intelligence can sift through mountains of healthcare data in real time, letting doctors make better, quicker estimates. Machine learning algorithms can detect trends that human analysts would overlook, which can lead to novel insights regarding the prevention and treatment of illness. Finally, artificial intelligence (AI) could help create a fairer healthcare system by supplementing ophthalmologists' knowledge with data on eye health extracted from images of the human eye.

- **Precision Medicine:** Personalised treatment regimens that are tailored to each patient's unique traits will be made possible through data analytics, which will be increasingly important as more information about genetic and environmental factors that impact health becomes available.

- **Telemedicine and Remote Monitoring:** The proliferation of wearable technology and telemedicine is creating new data streams in the healthcare industry, which, when processed, can offer continuous, real-time insights into patients' health. Data analytics will eventually enable healthcare providers to remotely monitor their patients, spot any health issues, and take action before they get worse.

- **Population Health Management:** The utilisation of data analytics will persist in revealing population-level health patterns and disparities, enabling healthcare organisations to execute focused interventions and enhance community-level health outcomes.

When properly applied, data analytics may significantly enhance the delivery of healthcare by making it more patient-centered, preventative, and economical. Given the growing reliance of healthcare systems on data-driven insights, the importance of data analytics in selecting the course of patient treatment will only grow in the future.

3.3 Compliance with Healthcare Regulations

The purpose of healthcare regulatory compliance requirements is to safeguard patients from damage and guarantee the provision of top-notch healthcare services. Healthcare professionals can better ensure patient safety by adhering to rules. These rules emphasise the importance of precise record-keeping, infection control, equipment sterilisation, and proper prescription administration (World Health Organization et al., 2018).

This research will explore the complexities of healthcare regulatory compliance, highlighting its importance in the industry and how Scrub can help your organisation achieve compliance.

In the healthcare industry, what is regulatory compliance?

Ensuring that healthcare organisations, professionals, and entities adhere to all rules and regulations established by governmental agencies and regulatory organisations is known as regulatory compliance in the healthcare industry. This is to guarantee that patients receive ethical, superior, and safe medical care.

Figure 3.1: Regulatory Compliance in Healthcare Covers

Source: - *(Automation, 2022)*

The delivery of health care services is governed by numerous regulations in order to safeguard patients, maintain the standards of the medical profession, and preserve the public's health. Within the healthcare industry, healthcare regulatory compliance encompasses several different areas, such as but not restricted to:

- **Patient care standards:** Laws outlined the standard measures that caregivers must adhere to prevent harm to the recipients. These guidelines include everything from diagnosis and care to patient rights and informed consent.

- **Data security and privacy:** HIPAA for example in the United States requires the protection of patient data and places a lot of demands on the healthcare organization to secure their data more securely.

- **Accreditation and certification:** Health systems seek accreditation from such bodies as The Joint Commission to prove their compliance to set standards of quality health care services. It is important to adhere to accreditation criteria.

- **Pharmaceutical and medical device regulations:** The Centre for Medicines Evaluation of the European Union EMEA and Food and Drug Administration FDA in USA are responsible for the regulation approval market and safety of drugs and medical devices.

- **Billing and coding compliance:** In the healthcare industry, organizations must bill and code appropriately and without fraud or billing abuses.

- **Workforce compliance:** The industry regulatory frameworks also dictate issues to do with competence and certification of healthcare professionals.

- **Ethical and legal practices:** Compliance also entails legal and ethical obligations, including those related to anti-kickback, research ethics, and conflicts of interest.

Regulatory compliance is of the utmost importance in the healthcare industry. The repercussions of noncompliance can be devastating, encompassing legal action, fines, licensure revocation, harm to one's reputation, and a decrease in patient safety. On the other side, healthcare rules are in place to keep the system running smoothly, safeguard patient privacy and rights, and guarantee high-quality care every time.

As a general rule, healthcare organizations that want to stay in compliance set up compliance procedures, hire compliance officers, audit and assess frequently, and educate and train their employees continuously. Compliance with rules in this complicated and heavily regulated industry requires constant monitoring of new legislation and adjustments to the healthcare system.

How is the regulatory landscape in healthcare at present?

The United States has some of the world's most stringent healthcare regulations. To guarantee the security of "public health information" (PHI) and its limited sharing to qualified parties, it incorporates checks at multiple levels. A patient's ability to have faith in the treatment they get is bolstered by the regulatory environment. Now we'll take a look at a few parts that make up the US healthcare system's regulatory structure.

Key regulatory bodies in healthcare

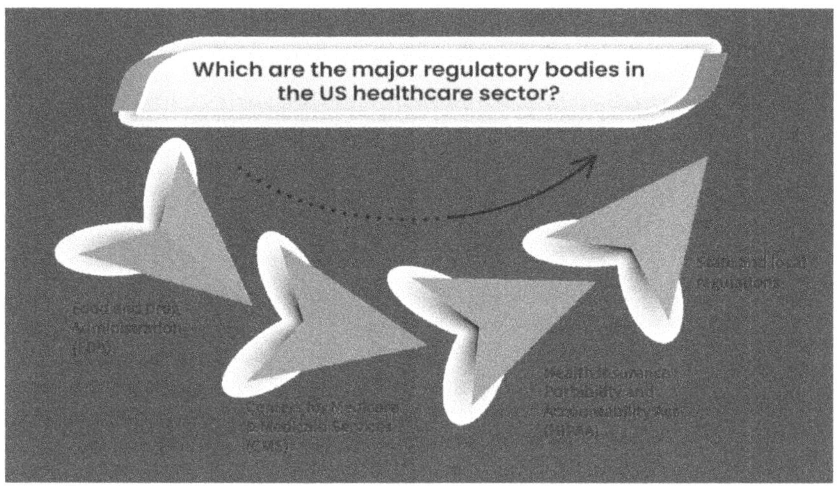

Figure 3.2: The major regulatory bodies in the US healthcare sector?

Source: - *(Automation, 2022)*

The following regulations guide the healthcare industry in the US at present.

1. **Food and Drug Administration (FDA):**

   The Food and Drug Administration (FDA) is a well-known US regulatory body that checks the legitimacy and effectiveness of foods, cosmetics, medical devices, and pharmaceuticals to protect the general population. It checks and certifies new health tech and pharmaceuticals, keeps an eye on product recalls, and makes sure no one gets in trouble for deceitful advertising or contamination.

2. **Centers for Medicare & Medicaid Services (CMS):**

   In the United States, CMS is in charge of the Medicare and Medicaid programs, which offer healthcare coverage to millions of Americans. It sets guidelines for healthcare providers taking part in these programs, monitors reimbursement rates, and makes sure that billing and coding rules are followed.

3. **Health Insurance Portability and Accountability Act (HIPAA):**

   Protecting the privacy of people's medical records is the goal of the "Health Insurance Portability and Accountability Act" (HIPAA). It requires healthcare organisations and their business partners to follow specific privacy and security guidelines for electronic health records (EHRs) to avoid unauthorised access or disclosure of patient data.

4. **State and local regulations:**

   Healthcare providers are not only subject to federal restrictions, but also to often-varied state and municipal requirements. Questions of scope of practice, healthcare facility standards, and license are all addressed in these rules. There is usually a separate health department or regulatory agency at the state level.

Evolving regulatory trends in healthcare

The regulatory landscape is subject to constant change due to the dynamic nature of healthcare and rapid advancements in technology. To maintain compliance, deliver high-quality care, and adjust to the changing healthcare landscape, healthcare organizations and professionals must stay updated about these developing rules and trends.

1. **Telehealth and virtual care compliance**

   There are new possibilities and possibilities for regulation brought about by the growth of telehealth and other forms of virtual care. When providing treatment remotely, healthcare practitioners face challenges related to licensure, reimbursement policies, and consumer privacy. To keep patients safe and provide high-quality care as telehealth expands, regulatory agencies are constantly changing to meet the needs of the industry.

2. Data privacy and security

The personal information of patients is seriously at risk from cyberattacks and data breaches. Regulators are strengthening security and data privacy regulations in an effort to address these problems. To protect patient data, it is essential to adhere to regulations such as the "General Data Protection Regulation" (GDPR) in Europe and the "Health Information Trust Alliance" (HITRUST) in the US.

3. Value-based care initiatives

The value-based care model is replacing the fee-for-service approach, which is altering healthcare delivery and reimbursement. The goal of new regulatory frameworks is to encourage providers to priorities outcomes and efficient use of healthcare funds. Care coordination, performance evaluation, and quality reporting are all part of these programs' compliance requirements.

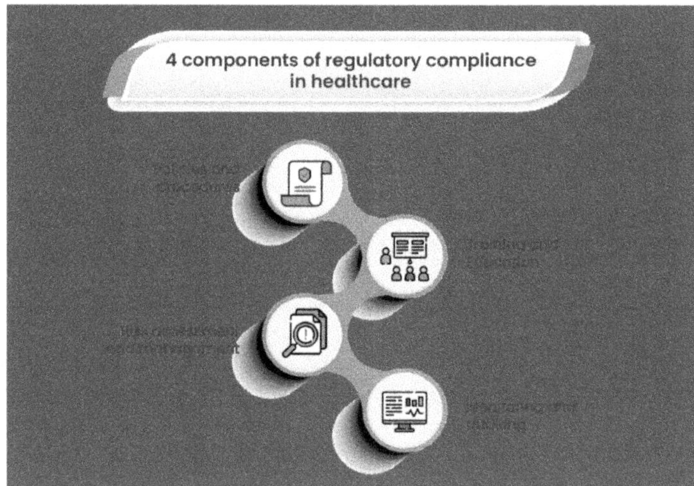

Figure 3.3: Core components of regulatory compliance in healthcare
Source: *- (Automation, 2022)*

The domain of healthcare regulatory compliance can be separated into four main categories. Join me as I enumerate every regulatory healthcare compliance that could exist.

A. Policies and procedures

Policies and procedures fall under two subcategories.

1. **Developing effective policies**

   Regulatory compliance in healthcare is built upon effective rules and procedures. Healthcare organizations are obligated to establish policies that are in line with regulatory standards and are clear, thorough, and current. Important parts of healthcare operations such as patient care, data security, and billing must be defined.

2. **Ensuring proper documentation**

   It is critical to document policies and procedures correctly. The development, revision, and distribution of policies should be documented by healthcare practitioners. Staff members can resort to documentation as a guide and it helps prove compliance efforts.

B. Training and Education

Educating and training staff is a crucial component of healthcare regulatory compliance. Rather than being a one-and-done task, it is a continuing practice.

1. **Employee training**

   It is recommended for healthcare organizations to embark on offering broad compliance education for all divisions especially the clinical centers, administrative departments, and all other help employees. It should be noted that awareness and understanding of regulatory, policy related, and procedural requirements must be a part of training.

2. Ongoing education

This compliance is an area that is dynamic, and over a period of time, rules change. Thus, staff members are required to be updated with the current requirements for compliance as well as the best practices. Updated multiparty training sessions, workshops, and reference materials assist in enforcing compliance as a culture.

C. Risk assessment and management

There are risks associated with every activity. According to the HIPAA Journal, there were 11 recorded healthcare data breaches in 2022 that exposed over a million records, and another 14 that affected over 500,000 records. Because of this, risk evaluations and management are crucial components (Murray-Watson, 2023).

1. Identifying compliance risks

Healthcare organizations should regularly carry out a risk analysis to know urgent compliance risks. This involves assessing procedures, guidelines or anything else the organisation and it's environment that might lead to compliance issues. Risk assessments are useful in a sense that they indicate where changes need to be made.

2. Mitigation strategies

When compliance risks have been identified in an organization, it is important that the healthcare organization put in place measures to manage such risks. This may include production flow improvements or procedures, managerial, organizational or administrative changes, or training on how to overcome some of the risks. Preventative measures are meant to minimize the possibility of non- compliance.

D. Monitoring and auditing

An important step is routine audits and monitoring. There are two smaller steps in it.

1. **Regular audits and self-assessments**

   To ascertain their degree of compliance with policy and the law, healthcare organisations are required to conduct internal audits and self-evaluations. Compliance audits maybe done by compliance officers or special compliance departments. These assessments assist in enlightening program administrators and the enforcing organizations about the existing and emerging compliance issues and issues that require redress.

2. **Corrective action plans**

If non-compliance is present then corrective action plan should be written to eradicate the deficiencies found at the health care organizations. These plans cover steps to address problems, steps to avoid repetition of the problem and steps to maintain and sustain compliance. It is common practice to determine corrective actions that may well include policy amendments, extra training, or process alterations.

Healthcare regulations must be approached with more than a reactive assessment of the protocols, polices, educational continuation, and risk management. Concentrating on these elements of compliance, healthcare organizations can more easily establish the culture that exclude many possible violations of the rules, ensure more responsible and more effective patients' treatment, as well as maintain the reputation among the official regulators and the general public.

Technology-based solutions for healthcare regulations

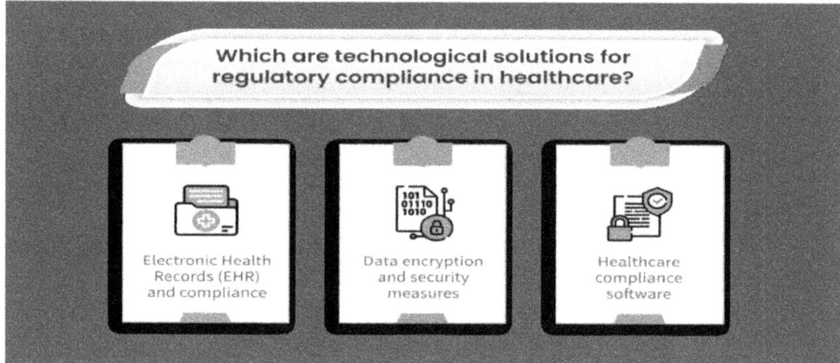

Figure 3.4: Technological solutions for regulatory compliance in healthcare

Source: - *(Automation, 2022)*

There are many solutions for regulatory compliance in healthcare. Here are some of them:

A. Electronic Health Records (EHR) and compliance

The preservation of patient health information is another area where EHRs are crucial for healthcare compliance. EHR systems support compliance in the following ways:

- **Data privacy and security:** EHRs are designed to meet legal standards for data security and privacy, including the United States' Health Insurance Portability and Accountability Act of 1996. They include measures like audit trails, user authentication, and access control to prevent patient data from being compromised by hackers or dishonest people.

- **Documentation compliance:** Medical history, treatment plans, and progress notes are just a few of the essential and comprehensive patient records that are maintained when EHRs are used. Addressing compliance concerns around patient care and billing requires this paperwork.

- **Automated compliance checks:** Current EHR systems for the most part include compliance checks and alerts to assist the clinician in following clinical and billing protocols. Such systems can alert possible problems like interaction between two drugs and missing documentation in real time.

- **Interoperability:** EHRs are used to transfer patient information among the clinics and other facilities for better organized and enhanced patient care and to abide by data-interchange rules.

B. Data encryption and security measures

Healthcare compliance requires data encryption and security safeguards, especially when it comes to protecting patient information:

- **Data encryption:** Implementing secure data encryption for the patient's ease is also important in that it secures the information in both storage and movement phases. Modern cryptographic techniques like AES are used as the security standard when passing the message.

- **Access controls:** The measures of access controls guarantee that within organization only the individuals with permission can get the information of patients, and other crucial data. RBAC systems assist in restricting users from what they need to do in their respective positions within an organization.

- **Regular security audits:** There is always need for Healthcare organizations to conduct routine security audits in order to pinpoint areas of insecurity and to determine the adequacy of implemented security measures respectively. Clear examples are penetration testing and vulnerability assessment, which can be performed to ascertain the vulnerabilities in the systems.

- **Data backup and recovery:** Any information back-up or disaster recovery measures are compulsory in meeting the requirement. Maintaining patient information and consistently having a strategy to recover data in the event of the mishap or theft are vital components of compliance.

- **Security awareness training:** There is a need for the employees to be trained on how they should exercise caution not to be part of a security threat. This is including awareness of common scams especially phishing and a good password management system.

C. Healthcare compliance software

Software that helps other organisations successfully track the status of their compliance can also be referred to as healthcare compliance software. Important attributes and advantages consist of:

- **Policy management:** These tools are helpful in creating policies and procedures, disseminating them, and making any necessary modifications. These are a few of the elements that frequently make up paper management and module version control systems.

- **Training and certification tracking:** Organisations may keep track of and record the training that staff members receive, their certifications, and the prerequisites for their further education and compliance thanks to this healthcare program.

- **Audit and reporting:** These tools make communications of internal audits, self-assessment, and compliance reports much easier. They are able to produce other reports showing that part that is not in compliance and other one showing the remedial measures.

- **Risk assessment:** Many compliance programs are comprised of risk assessment tools to locate and rank compliance risks with the intention of assessing likely problems.

- **Incident reporting:** Usually, incident reporting is integrated into healthcare compliance software so that staff members can safely and discreetly report compliance problems, privacy violations, or other incidents.

Scrut is among the best applications for managing regulatory compliance. You will learn how Scrub can help you with healthcare compliance in the next section. However, it will be important to research some of the problems that can arise when attempting to create a robust system of health care compliance before moving forward.

Healthcare compliance issues and solutions

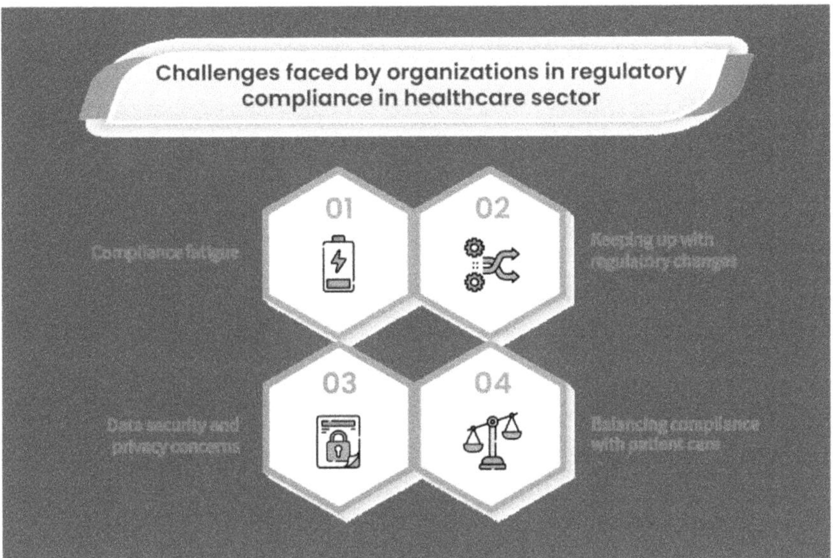

Figure 3.5: Challenges faced by organizations in regulatory compliance in healthcare sector

Source: - *(Automation, 2022)*

Here are a number of the challenges – the effect that the above challenges can have on the organisation – and how to handle these.

A. Compliance fatigue

- **Challenge:** Compliance fatigue is the phenomenon in which concern and the level of attention to various rules and regulations is decreasing, caused by the overwhelming number of rules among health care workers and companies.

- **Impact:** Sometimes, people and organizations reach a state of mere tolerance of compliance requirements and, therefore, may develop compliance failures which are acts of noncompliance with essential compliance standards. This can lead to such consequence as increased violations rate, jeopardized patient safety, legal and financial consequences.

- **Mitigation:** To avoid compliance fatigue among the healthcare organizations it is recommended that some interventions like conducting training and awareness sessions from time to time, providing adequate information on the significance of compliance, and the use of information technology in overseeing the compliance activities as well as being integrated into working plans.

How can Scrut help? Scrut makes the employee infosec training fully automated with a default 30-Infosicherheits-Distanzierung that is professionally developed. This training provides staff members with all information necessary to get aware of possible risks, avoid escape and build a secure position. The platform also provides employees an easy approach to view policies without difficulty, assess notifications, and appreciate security procedures at the same time.

B. Keeping up with regulatory changes

- **Challenge:** Hailed care regulations are the ever-changing phenomenon, and it always has a tendency to undergo amendments at federal and state & local levels. Many a time it is difficult to keep abreast with such changes.

- **Impact:** Not being able to produce goods and services at the speed of the changing regulations may expose you to compliance issues. Healthcare organizations may have policies and procedures in place that are outdated that puts the healthcare organization in a legal vulnerability.

- **Mitigation:** Thus, healthcare organizations should assign those people or teams that will be responsible for the tracking of the shifts in the regulations. This might involve becoming a member to regulatory newsletters, setting up to go to industry meetings, and having compliance software which brings updates. Compliance audits can also be used in helping in understanding which aspects of the compliance management process need some fine-tuning because of the change in regulations.

How can Scrut help? With this platform, you can intercede with your auditors and even directly engage with them. Through all the implementing policies, controls and proofs accumulated in one place, audits do not take much time and efforts.

Scrut also fits into your environment as an integrated solution for handling evidence collect as well as for managing compliance assignments where you build, assign, monitor and track duties. To guarantee the automation of the collection of human evidence, it interfaces with a variety of application landscapes, including HRMS, endpoint management, and other technologies. The program handles the most important hazards and effortlessly extracts evidence from over 70 applications.

Nevertheless, compliance audits are easy to conduct and all the necessary data is easily accessible because you maintain all of your rules, procedures, controls, proof, and documentation within the Scrut platform.

C. Data security and privacy concerns

- **Challenge:** As will be discussed in this study, such organizations deal with huge amounts of confidential patient information, which makes them a lucrative target for hackers and data criminals. Preserving the organization's comprehensive data protection and data privacy policies is always difficult.

- **Impact:** Databases combined with security risks that can lead to a range of losses such money, reputation, fines and penalties, as well are patient loyalty. To avert these risks healthcare organizations has to take data security as central.

- **Mitigation:** Use encryption of data, the usage of access controls, constant security audits and employee education about cybersecurity. HIPAA for instance, is a member of data privacy laws that must be fully observed when handling patient's details.

How can Scrut help? Scrut Risk Management: The platform is intended to act as a venue for identifying and evaluating IT and cyber risks, as well as for putting suitable countermeasures in place.

It gives organizations the understanding of controlling risks they require to avoid them and informs management on risks affecting high-priority strategic objectives. Due to the feature of scoring techniques, using risk scores provided by the experts and applying automated workflow, the users of the platform can easily create new IT risk programs within a short time and the platform can help prevent, control and mitigate the risks.

Among Scrut's distinctive features is the ability to use the tool as a risk register. A risk registry is a list of the dangers to which an organisation is exposed, including the risk's specifics, likelihood, and consequences.

The advantage of Scrut Risk Management is that every pre-loaded risk in the library, a new one created, the method of treatment, the workflows for mitigation and the accountability all exist in one central place.

D. Balancing compliance with patient care

- **Challenge:** The caregivers therefore often tread on a thin line between being very keen on compliance measures on one end, and the mission of delivering quality healthcare service on the other. Reasons why compliance can sometimes be a drawback include but are not limited to the following; In the context of compliance, they sometimes focus a lot of attention on efficiency as well as the provider-patient communication.

- **Impact:** Adherence to bureaucratic-oriented policies without considering its implications to patients will reduce patient satisfaction, and quality of services and increase staff turnover.

- **Mitigation:** To achieve the right balance on compliance, there is a need to communicate compliance with care delivery expectations, integrate compliance to the least extent possible into care delivery models and make organization-wide compliance prevalent. Clinician engagement in compliance decision-making and use of compliance as a strategy for realizing patient safety and quality can achieve this balance effectively.

How Scrut can help? Scrut changes the rules of the compliance game by cutting 75% of manual work time, increasing the level of accountability and the pace of InfoSec tasks. This platform provides a better solution to this challenge by reducing time spent on handling manual procedures and at the same time provides real time on the success of your programs.

With Scrut smartGRC, you stay informed about the overall status of your GRC program, as depicted in the accompanying screenshot:

Scrut performs all compliance duties without any delegation or decentralization of any of its tasks. If you're new to building your compliance program, we provide a library of more than 50 policies that you can tailor and review with our experienced in-house infosec specialists.

Combined, healthcare regulatory compliance is one of the fundamental tenets of the healthcare sector, that preserves the interests of patients, and maintains the standards of professional and business ethics as well as the credibility of the industry. Conformity is more than a legal requirement – it is a responsibility to deliver only safe and quality care. The rules of the healthcare industry are changing all the time and this is particularly evident in America. Healthcare organizations to analyze new and changing regulations and those working in healthcare organizations must also ensure they are up to date with current and current changes. It is not simply the fact of compliance but whether this compliance is carried out as part of a culture that emphasises duties and obligations.

In today's environment of protected healthcare information and strict rules on data protection, the highly developed services like Scrut are especially valuable. It is a practical source that provides options for improving compliance activities with considered elements and factors due to the ability to increase their efficiency by automating them. Starting from policy management up to risk assessment, features offered by Scrut can help prevent leakage of valuable patient information.

3.4 Integrating Electronic Health Records (EHR) with Cloud BI

The adoption of EHR and cloud BI is an innovative solution that enables an organization to fully realize the value of patient data to inform important decisions that benefit a healthcare organization, its operations, and the patients it serves. EHR systems include all patient data in such aspects as medical history, treatment plans and details, laboratory results, and medication. Nevertheless, such data is stored in various systems, and it is challenging to pull valuable information without having a single integrated platform. This is overcome by cloud BI systems since they consolidate data from various EHR platforms and this will provide the professionals with a source of truth that was not present before, from where they can analyze and report (Wang et al., 2018).

Establishing access relies on the user's identification being reliably validated before they may access protected health information (PHI). Historically, healthcare providers have operated independently, with each building their own data storage silos and security frameworks to access those stores. As shown in Figure 3.6, the conventional wisdom on electronic health record (EHR) system authentication holds that the validating credentials are held within the application itself. Figure 3.6 indicates that the credential store or identity provider is typically embedded within the program itself in older systems. Eliminating this reliance is a primary goal of this research. To put it more simply, this study suggests that electronic health record apps should not be limited to using credentials stored in the local EHR database for credential validation; they should be able to access other identity stores as well. Consequently, healthcare providers can delegate the authentication challenge to other manufacturers and organisations who have already invested heavily in this field while they focus on electronic access to their EHR systems. (Norgeot et al., 2020).

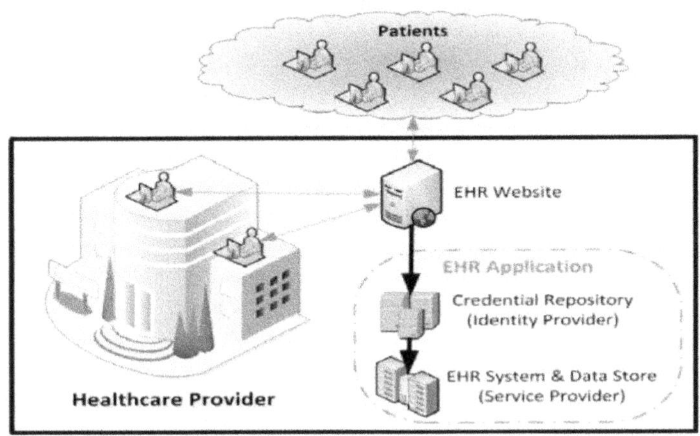

Figure 3.6: Traditional Electronic Health Record (EHR) Access Model

Source: - *(Coats & Acharya, 2014)*

There are many benefits to be gained from Cloud BI integration with EHRs. As it is real-time data analysis it is easier for the clinicians to follow up on the patient progress and the health trends in a given region. As such, through concrete visualizations of data, including dashboards and dynamic reports, healthcare providers can hold important and seemingly simple KPIs such as readmission rates, average length of stay, and treatment effectiveness. Also, this integration can help in population health management where data is collected from different groups of patients to determine tendencies, risk indicators, and potential focus points for interventions.

By using the ability to separate the authentication process from the EHR application, this research provides a framework that enables a single EHR system to be configured to employ an arbitrary number of authentication systems, as seen in Figure 3.7. Any trustworthy identity provider can perform the authentication event in this architecture.

Commercial organisations that already have a relationship with people are the best providers of identity. Verizon, Comcast, AT&T, Google, Yahoo, and Microsoft are a few of the more well-known contenders. By utilising pre-existing cloud credentials, healthcare organisations may efficiently offload the burden of creating and maintaining usernames and passwords while also offering their patients a familiar user experience. The cloud's current identity suppliers spend billions of dollars annually on usability. Making their services accessible and visible over the Internet is at the heart of many of their usability initiatives. The cloud's core authentication features, such retrieving a username or changing a password, are always being enhanced. Actually, one of the biggest technical support problems for businesses is changing passwords, as is covered in the Results section below. This paradigm outsources this support issue by allowing healthcare providers to simply leverage the cloud's existing investment instead of trying to reproduce it. Additionally, many cloud identity providers offer their services to organisations at no cost, except for the man-hours required to set up the integration.

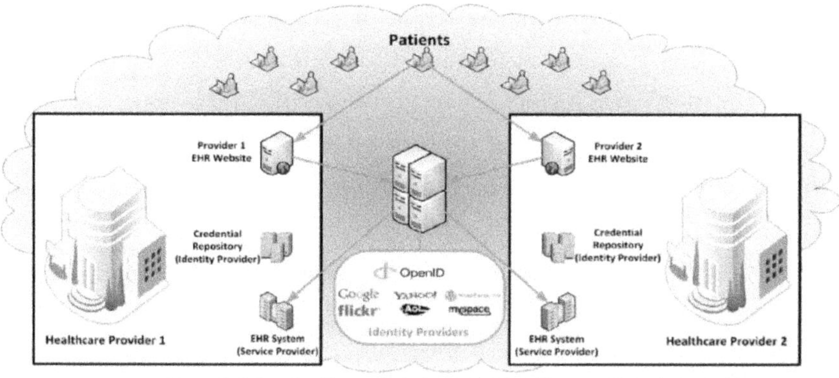

Figure 3.7: Cloud-Based Federated Electronic Health Record (EHR) Access Model

Source: - *(Coats & Acharya, 2014)*

One of the main advantages of using EHR in combination with cloud BI is the ability to increase the availability of data and work with it collectively. Cloud platforms enable only the authorized people in different centers and departments to access and analyze the patient data, encouraging the cross- department working. For example, a hospital network that has many hospitals could apply the use of a cloud BI system to combine patient data, and it becomes easy for the specialists to coordinate with other professionals or share information to and from other hospitals. Besides, it enhances and organizes the actions of care delivery, minimizing duplication of services and thus saving costs and utilization of resources.

Nevertheless, there are few technical and regulatory issues to be discussed to achieve the goal of integration. This is especially the case for data exchange which remains one of the biggest issues because EHR systems utilize various data formats and standards. System integration for patient data transfer between platforms can be streamlined with the use of HL7/FHIR (Health Level 7 and Fast Healthcare Interoperability Resources) standards. Additionally, strong laws, including the Health Insurance Portability and Accountability Act (HIPAA) in the US, which requires tight data privacy standards, control patient data access to healthcare companies. Cloud BI solution architecture needs to be sufficiently secure to safeguard such private patient genetic information while also adhering to legal requirements.

Another important consideration while adopting EHR together with the cloud BI is scalability. Hospitals and other healthcare centers receive large volumes of data daily, including, among the patients' electronic medical records, diagnostic images, and data from wearable health devices. Cloud BI platforms must be able to scale, to handle big data workloads without compromising on the speed and quality of data. Using cloud-native platforms like data lake and distributed computing can ensure no scalability issues as the amount of data expands.

In sum-up, EHR with cloud BI systems enables the healthcare organization to make effective BI applications for enhancing the clinical decision-making process, enhancing patient care delivery, and efficient organizational operations. It is a tool for the coordinated access to trustworthy data where data integration breaks the barriers of data silos and supports the real-time access of data as means to an overarching proactive approach to advance the healthcare services while conforming to all standards set by the healthcare industry.

3.5 Predictive Analytics for Disease Management

What Are Predictive Analytics in Healthcare

Predictive analytics is a branch of data analytics that primarily makes use of machine learning, data mining, modelling, and artificial intelligence techniques. It is used to evaluate both historical and present data in order to predict the future (Lee et al., 2022).

Predictive analytics, as used in the healthcare sector, is the analysis of historical and current clinical and operational data that helps physicians predict trends, pinpoint areas for improvement, and even stop the spread of disease.

Data about health collected through medical and administrative records, health surveys, patient and disease registries, claims-based datasets, and electronic health records is referred to as healthcare data. Healthcare analytics may help everyone working in the healthcare industry—hospitals, healthcare organisations, physicians, psychologists, chemists, pharmaceutical companies, and even stakeholders—provide better care.

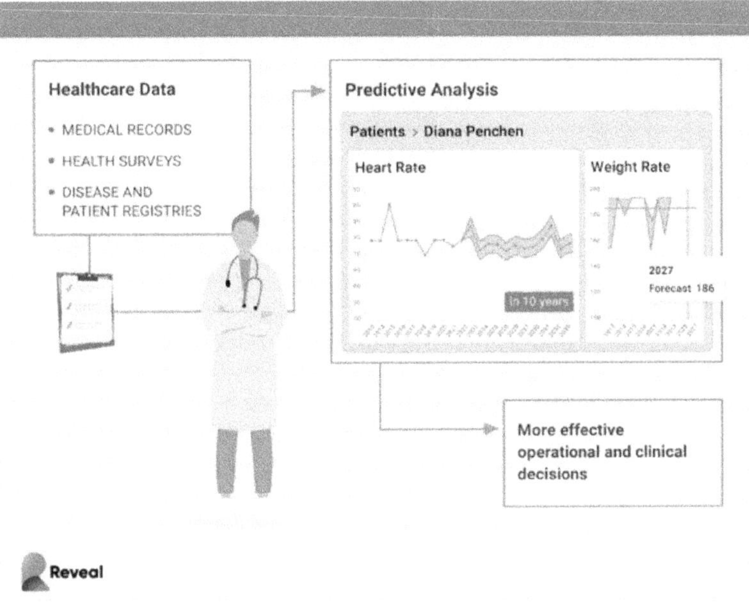

Figure 3.8 Predictive Analytics in Healthcare

Source: *- (Petrova, 2024)*

Healthcare Applications of Predictive Analytics

Although the health care sector generates enormous volumes of data, it struggles to transform that data into insights that could benefit patients. Utilizing data analytics in all aspects of patient care and operational management is the aim of the healthcare sector. It is used to predict disease outbreaks, improve patient care, investigate methods to reduce treatment costs, and much more. Analytics can help healthcare businesses streamline internal procedures, maximize resource utilization, and improve care team coordination and efficiency at the corporate level.

The ability of data analytics to transform raw medical data into actionable insights has a significant impact on the following healthcare domains:

- Discovery of new drugs

- Automation of hospital administrative processes

- Prediction and prevention of diseases

- Development of new treatments

- High success rates of surgeries and medications

- Quicker and more accurate diagnosis of medical conditions

- Clinical decision support

- Clinical research

- More accurate calculation of health insurance rates

Benefits of Predictive Analytics in Healthcare

As technology develops, analytics can have a big impact on the healthcare industry. AI and machine learning techniques can use data to diagnose diseases and determine the best course of treatment for each patient's particular circumstances, among many other things. Healthcare businesses can benefit from predictive analytics in the following important ways:

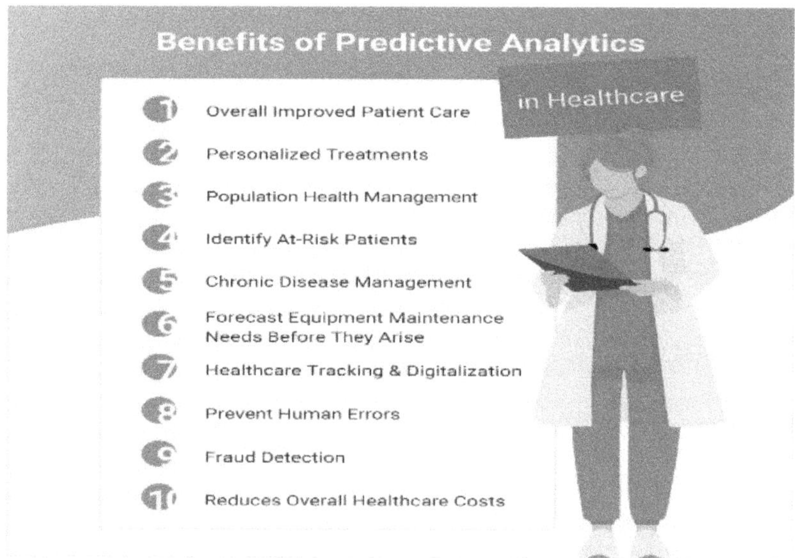

Figure 3.9: Benefits of Predictive Analytics in Healthcare

Source: - *(Petrova, 2024)*

- **Improved Patient Care:**

  The biggest benefit predictive analytics provides to the healthcare industry is access to all types of data, including demographics, economics, comorbidities, and medical history. Physicians and other medical professionals can use all of this information to inform their decisions. Overall, making smarter, more data-driven, and educated judgements leads to better patient care.

  For example, predictive analytics is utilized to improve patient outcomes. In order to provide insights into the best treatment strategies for each patient, machine learning algorithms can be taught by looking at the data and outcomes of prior patients.

- **Personalized Treatments:**

  Historically, medicine has taken a one-size-fits-all approach. Instead of treating individual patients, medications and treatments have been provided based on statistics of a large population and scant information. On the other hand, because they are able to diagnose patients more precisely, doctors are able to choose the best course of action for each patient's particular health condition.

- **Population Health Management:**

  Individuals are not the only ones who can use predictive analytics. Healthcare organisations can use it to control population health as well. Analytics can be used to find similar patients in a population cohort based on information about a patient's health history, medications, and conditions. It can also help identify cohorts exposed to a possible pandemic of disease. Medical professionals can start thinking about treatments immediately in such a scenario, improving the chances that patients will live.

- **Identify At-Risk Patients:**

  Predictive analytics in healthcare can detect high-risk individuals and start early innervations to avoid more serious problems. For example, it can identify which patients with cardiovascular disease are more at risk of hospitalization based on medication adherence and age-coexisting chronic diseases. By employing projections on the likelihood of disease and chronic illness, doctors and healthcare organizations can provide proactive care to at-risk patients rather than waiting for them to come in for a standard checkup.

In addition to the chronically ill patients, additional categories at risk include the elderly and individuals recently discharged from the hospital after invasive procedures.

- **Chronic Disease Management:**

  In the US, chronic illnesses are the primary causes of mortality and disability and account for $3.5 trillion in yearly health expenditures. 75% of healthcare cost is attributed to five chronic diseases: kidney disease, diabetes, obesity, cardiovascular disease, and cancer.

  In order to effectively treat chronic diseases, healthcare providers must be able to both prevent and control their progression. Chronic illnesses are difficult to avoid and manage, though. Healthcare professionals can use predictive analytics to make timely, fact-based decisions that will result in more effective treatments and lower patient expenses.

- **Forecast Equipment Maintenance Needs Before They Arise:**

  Other industries, such as manufacturing and telecommunications, have long used predictive analytics to foresee maintenance needs. The healthcare industry may benefit from the same kind of prognostics. Over time, machine parts wear out or deteriorate. For example, by analysing sensor data from an MRI machine, predictive analytics can predict component failures and when a component will need to be replaced. By using this data, hospitals may schedule maintenance for when the machine is not in use, which minimises disruptions to patient and care team operations.

- **Healthcare Tracking & Digitalization:**

  The economic and technological enabler of health care services erases entirely the context of the relationships between patients and health professionals. Today, it is possible to wear some sort of device which would monitor one's health and his/her bodily performance at any given time through his/her mobile phone. For instance, it enables diabetics to check up the increase in blood sugar at any one time without having to use a pin prick.

- **Prevent Human Errors:**

  Potential catastrophic consequences of human errors are fatal in relation to healthcare. Fortunately, with real time and relevant information for medical practitioners to act on, data can alert one to an error and possibly help avoid a fatal decision.

- **Fraud Detection:**

  However, that is not so surprising since fraud is present in a number of spheres, including healthcare. Fraudulent healthcare schemes come in many forms: such things as people getting prescription pills that have either been subsidized or fully-covered only to turn around and sell it out on the black market; getting compensated for a non-covered service but billing it under a covered service; incorrect medical record alteration; inaccurate reporting of diagnosable conditions or treatments, among others.

  It is noteworthy that predictive analysis can pick on some anomalies which point to such fraudulent actions, therefore early detection.

- **Reduces Overall Healthcare Costs:**

  It is also possible to apply predictive analytics in decreasing healthcare costs. They can be utilized to shave down costs since patients will not be admitted to hospital where it is not necessary, to control hospital costs of drugs and other essential commodities and to determine future requirements for nurses and other employees.

Data mining application in chronic illness care may change the face of medical treatment by moving from the treatment to prevention models. Diagnostic tools, EHR systems, and wearable health devices that collect enormous data on patients' health can predict disease risks and create prevention and individualized care plans before the onset of the disease. Although this approach has the added advantage of enhancing patient health outcomes, it has the additional benefit of decreasing the costs of healthcare since complications and readmissions are minimized. The application of cloud BI capabilities to predictive analytics adds even more perspective to this approach, providing, for example, real-time analyses, virtually unlimited processing capabilities, and collaborating security for healthcare networks. Finally, the use of the predictive analytics on illness management promotes a detailed oriented care delivery system which is patient centered and focused more on prevention as well as early detection.

3.6 Challenges and Solutions in Healthcare Cloud BI

Challenges come with adoption and implementation of cloud-based BI especially in the healthcare sector because of data environment, legal issues and security issues. This has been found to be one of the main concerns as any data collected should be protected and should meet the act HIPAA. Therefore, healthcare organizations should use encryption and control techniques like role-based access controls and audit trails to safeguard the identified comprehensive patient information in order to prevent unauthorized access and breaches. (Al-marsy et al., 2021).

Another challenge that has been identified is the gateway between "Electronic Health Record" (EHR) systems and other data sources that are either disparate or come from different vendors as well. Most healthcare organizations employ multiple applications that employ dissimilar formats and standards for storing data, which results in the creation of silos. To this, the healthcare organizations can use Universal data exchange formats such as HL7 and FHIR, which facilitates data interoperability between many systems. Some Cloud BI platforms can even help in taking much of this complexity away by centralizing and harmonizing information from different sources for multi-Cloud analysis.

In healthcare BI there is also an issue of data quality and data accuracy. When data is incomplete, outdated or inconsistent it leads to wrong conclusions and ultimately wrong clinical decisions. Adoption of data governance and use of automated validators helps eliminates wrong data from the database. It is also necessary to conduct data audits periodically and provide data entry staff with training on proper data entry procedures that will ensure the quality of data in the organization is high (Karkouch et al., 2016).

It is increasingly important to consider scalability and performance since the amounts of healthcare data are rising rapidly. Cloud BI platforms must also be able to accommodate the big data sets, derived from EHRs, diagnostic imaging systems, remote patient monitoring devices and other sources of health data. Shading, data partition, and data warehouse, and other scalable works using cloud computing enable high-speed data processing at the same time with large data loads for real-time analysis.

There is always the issue of high costs; this is especially so since cloud BI implementation calls for high initial capital investment and continuously charging for subscription services. To overcome this, the healthcare organizations should consider the following factors while selecting a BI vendor – the cost structure, the flexibility of scaling up and down, as well as the support services offered. Most cloud BI solutions are customizable to an organization's size and data requirements and thereby costs are well controlled.

Finally, user adoption and technical expertise will continue to be a huge impediment. The decision makers in the health care facilities may not possess adequate expertise to properly utilize cloud BI tools. To counteract this, it is suggested that organizations focus on key and easy to use BI platforms, with clear and concise interfaces and ensure an extensive training of their users. If lower-echelon end users are involved right from the implementation process and are provided with constant support, there is a high likelihood the implementation will be successful and cloud BI seen to offer maximum benefits in the health sector.

Thus, solving these issues with the help of corresponding tactics, healthcare organizations will be able to use cloud BI in all its potential for providing the best decisions based on the data, increasing the quality of treatment and the overall management of the organization's processes.

3.7 Case Study: Cloud BI Implementation in a Hospital Network

A large hospital group with many facilities in various areas had severe problems with data consolidation, sharing, and analysis because of isolated data repositories. It is also important to note that data was fragmented across many local systems such as EHR, billing systems, inventory data bases, etc., and therefore it was cumbersome to gather coherent and meaningful information. Decision-makers and other medical staff had limited visibility into such data to support timely decision-making. Moreover, the overall hospital system had to adhere to the regulations of confidentiality that include HIPAA and GDPR when it comes to dealing with the patients' data. Another set of issues was inappropriately manual reporting processes which impacted the overall efficiency and left the network without the possibility to utilize more enhanced approaches to analytics for further improvement of patient treatment outcomes as well as general network functioning (Ahmed & Hussan, 2018).

In order to overcome these challenges, the hospital network has decided to integrate Cloud Business Intelligence (BI) solution which will involve capturing data from various branches and put it into a single platform. The initial goal of the project was to achieve the real-time data access through dashboards, improve security and compliance with the data and apply predictive analytics to the resources and patients' needs.

The chosen solution entails using the Microsoft Azure for Cloud security and scalability in data storage, Power BI for interactive reporting and data visualization and data integration using Azure Data Factories. Safeguards that were implemented were: encryption; client input and output controls; access controls: and authentication for customer information protection (Chen et al., 2012).

The implementation process was done in the form of a sequence of sub-processes, including the analysis of the current state of the data architecture, transferring the data to the cloud securely. The next phase followed data transfer was data integration and a lot of testing to get it right. Hospital network also invested a significant amount on staff training and change management in order to create awareness on right use of new BI tools. The last stage was the comprehensive introduction of the Cloud BI system and its permanent improvement and control.

After the Cloud BI was put in place, the hospital network realized lots of improvements. Real time dashboards gave decision makers better access to timely data they needed to make better decisions regarding patient care and resources in the facilities. Implementation of automated reporting of time and attendance reduced the amount of repetitive manual work done by the staff to a great extent and HIPAA and GDPR compliance was improved due to additional security implementations. Also, realigning the infrastructure to the cloud reduced operational costs since there was little need for maintaining hardware facilities.

But as it was expected, the process of implementation was not without its pullbacks. A few staff members have heavily resisted to change of new systems while data quality posed challenges that needed to be cleared before integrating. While on the area of integration, compatibility with the legacy system was another challenge that could slow down the flow of data.

The lessons learnt from this case study include engaging stakeholders early, focusing on data protection and doing implementation in phases to cause less interference. Effective planning of the Cloud BI project, engaging technology partners and extensive training were considered to be the major success factors. Finally, the project evolved the data infrastructure of the hospital network for providing improved patient care and data-drive decision making for better operations across the network.

3.8 Chapter Summary

In this chapter the author examines the potential of cloud-based business intelligence (CBBI) and data analytics in healthcare organizations regarding increased patient satisfaction, organizational effectiveness and compliance with healthcare standards. They demonstrate how CBBI can consolidate data, improve data availability as well as incorporate it into more accurate predictive models for decision making purposes, all the while meeting regulatory compliance measures as detailed in the HIPAA standards. The chapter also discusses the different categories of healthcare data analytics which include predictive analytics, diagnostic analytics, descriptive analytics, and prescriptive analytics and their uses in treatment and patient handling, disease control and management of healthcare institutions. Also covers the topic of the given proposal about compliance, data security, and the interface of EHR with cloud BI for data sharing. The success of the CBBI implementation in the hospital network results from the enhanced reporting, decision making, and reduction of operational cost as well as communication with the stakeholders, staff training as well as addressing data security measures.

Multiple Choice Questions (MCQs)

1. **What is the primary goal of implementing Cloud BI in healthcare?**

   a. Reducing staffing levels

   b. Enhancing patient care through data-driven insights

   c. Minimizing hardware costs

   d. Replacing EHR systems

2. **Which of the following is an example of a healthcare regulation that must be complied with when using Cloud BI?**

   a. GDPR

   b. SOX

   c. HIPAA

   d. FLSA

3. **How does Cloud BI enhance patient care?**

   a. By automating surgical procedures

   b. By enabling advanced data analytics for personalized treatment

   c. By reducing the need for patient records

   d. By replacing healthcare professionals with AI

4. **What is the primary purpose of integrating Electronic Health Records (EHR) with Cloud BI?**

   a. To digitize paper records

   b. To improve access to real-time data and analytics

   c. To eliminate patient privacy concerns

   d. To standardize all healthcare practices globally

5. **Predictive analytics in healthcare is primarily used for:**

    a. Diagnosing diseases without clinical input

    b. Predicting and managing potential disease outbreaks

    c. Eliminating the need for medical imaging

    d. Standardizing treatment protocols across hospitals

6. **What is a significant challenge when implementing Cloud BI in healthcare?**

    a. Lack of healthcare regulations

    b. Ensuring data security and privacy

    c. Incompatibility of cloud platforms with EHR systems

    d. Excessive patient data

7. **Which of the following is NOT a benefit of Cloud BI in healthcare?**

    a. Real-time insights into patient care

    b. Enhanced compliance with data regulations

    c. Reduced costs associated with data storage

    d. Elimination of all human errors in treatment

8. **In the case study, what was a key benefit observed after implementing Cloud BI in a hospital network?**

    a. Elimination of administrative staff

    b. Improved decision-making through centralized data

    c. Replacing traditional IT infrastructure with blockchain

    d. Reducing patient visits

9. **What role does data visualization play in healthcare Cloud BI?**

    a. It replaces clinical reports with graphical summaries

    b. It helps stakeholders easily interpret complex data

    c. It generates automated diagnoses

    d. It removes the need for predictive analytics

10. **What is an effective solution for addressing data privacy challenges in Cloud BI?**

    a. Encrypting sensitive patient data

    b. Using outdated hardware for data storage

    c. Limiting access to analytics tools

    d. Storing data only on physical servers

Answer

1	2	3	4	5	6	7	8	9	10
b	c	b	b	b	b	d	b	b	a

04

CLOUD BI APPLICATIONS IN RETAIL

4.1 Chapter Overview

The retail Industry has been transforming currently with adoption of new business solutions such as Cloud Business Intelligence (Cloud BI). This chapter also looks into how Cloud BI impacts the flow of products in retail stores, refines customer experiences, and enriches a retailer's personalization plans by leveraging intelligence from data. The paper explains Cloud BI adoption for supply chain management and also discusses the concern of data security especially from the retail perspective. The chapter, supported by a real-life case, highlights how one retail chain managed to transition to Cloud BI and reach increased flexibility, better decision-making, and better consumer satisfaction within today's competitive environment.

4.2 Optimizing Inventory Management through Cloud BI

What is Cloud Inventory Management?

With the use of modern technology known as cloud inventory management, businesses can efficiently track, manage, and optimize their inventory in the cloud. With real-time visibility, automated workflows, and seamless integration, it improves supply chain efficiency overall, increases accuracy, and optimizes inventory operations (Akshay Badkar, 2023)

Figure 4.1: Inventory Management System

Source: *- (Education, 2024)*

How cloud inventory management works

Gone are the days of endless spreadsheets and manual inventory counts — cloud-based inventory management systems have revolutionized how businesses track and optimize their stock. Let's explore the inner workings of these systems and discover how they empower businesses to control inventory like never before.(Jonny Parker, 2024)

1. **Centralized database**

   Cloud-based inventory management systems gather data from various sources, including information about inventory levels, purchase orders, and raw material accounts. Many can even offer a breakdown of how your inventory is allocated and where it's located.

Imagine you have warehouses in Los Angeles, New York, and Denver. A cloud-based system consolidates inventory data from all three locations into a single dashboard, giving you an overview of stock. This is essential for accurate inventory tracking and informed decision-making about redistributing or reordering product.

2. Real-time visibility

A great system instantly reflects every transaction — sales, returns, new shipments, and more.

If a sudden surge in demand depletes a particular product, the system will alert you, allowing you to reorder before you run out. Setting reorder points is crucial for preventing stockouts, while the system gives you minute-by-minute data on what's selling and being received.

3. Analytics and reporting

The most effective cloud-based inventory management systems estimate demand and pinpoint areas for performance enhancement by examining past data, consumer behaviour, and market trends.

They help you create reports on your top-sellers, stock-dwell, and high point of sale periods where you make decisions that will directly impact your profits. These analytics and reporting are crucial not only for inventory planning that will generate the highest possible ROI.

4. Scalability

That is why your inventory management system has to evolve as your business expands. Cloud solutions on the other hand provides the option of either expanding it if the business grows or downsizing if the business shrinks. This results in designing new systems from scratch, or modifying and extending existing ones from adding a new distribution center to expanding in international markets, cloud-based systems are flexible enough to accommodate for change.

What are the Different Types of Cloud Inventory Management?

There are several varieties of cloud-based inventory management, such as just-in-time (JIT), distributed, centralised, and real-time inventory management.

Features of Cloud Inventory Management

A variety of strong features are available in cloud inventory management to help you optimized and streamline your inventory procedures. The most crucial aspects to take into account are as follows:

- Real-time visibility of inventory levels

- Centralized data storage and accessibility

- Streamlined inventory tracking and reporting

- Automated inventory replenishment and ordering

- Integration with other business systems

- Scalability to accommodate growing inventory needs

Benefits of Cloud Inventory Management

Cloud inventory management has several advantages for companies of all sizes. Businesses may improve accuracy, increase overall efficiency, and streamline inventory procedures by utilising cloud-based solutions. It makes it possible to improve inventory control, cut expenses, and boost customer satisfaction through automated tracking, real-time visibility into inventory levels, and smooth connection with other systems (shopify, 2024)

1. **Increased visibility across the organization:** Benefit from enhanced organisational visibility with cloud-based inventory management, which enables real-time tracking, efficient operations, and better

decision-making. Use the cloud's power to your advantage for the best inventory management.

2. **Greater accuracy and traceability:** Businesses can track inventory in real-time, minimize mistakes, improve order fulfilment, and increase supply chain visibility using cloud inventory management's increased accuracy and traceability.

3. **Saves time and money:** By offering real-time insight, optimizing stock levels, simplifying procedures, lowering human mistakes, and doing away with the need for expensive on-premises equipment, cloud inventory management saves time and money.

4. **Remote and real-time warehouse management:** Learn the revolutionary advantages of cloud inventory management, which makes it possible to operate warehouses remotely and in real time. Utilize this cutting-edge solution to improve accuracy, streamline processes, and obtain insightful knowledge.

5. **Streamlined supply chain management:** Take use of cloud-based inventory management to streamline your supply chain management. Boost performance by controlling the inventory with real-time visibility, increased accuracy, and positive change.

6. **Scales with the business:** In fact, one of the main benefits of cloud inventory management is its adaptability, which allows businesses to quickly grow or shrink their operations while knowing that they will always have a productive inventory system.

7. **Improves customer satisfaction:** Customer satisfaction is increased by cloud inventory management because it provides them with precise stock information, quick order processing, and real-time visibility, all of which lead to effective delivery and meeting customer demands.

Implementing Cloud Inventory Management

One way to define cloud inventory management is as the new technology for managing inventories on the cloud. Effective inventory control, improved performance, and up-to-date inventory information are all made possible with the aid of cloud solutions. Businesses may use cloud-based inventory management to revolutionise cost control, customer happiness, and inventory control, laying the groundwork for expansion and success in the present competitive market (Akshay Badkar, 2023).

1. **Evaluating and selecting the right cloud inventory management solution:** However, while using cloud inventory management, there are a few crucial procedures to adhere to. For increased efficiency, these requirements address scalability, integration, and real-time inventory views.

2. **Data migration and integration with existing systems:** High degrees of data transfer and integration with existing systems for cloud inventory management that is quick and trouble-free, leading to more accurate data and better inventory procedures.

3. **Training and onboarding of employees for successful adoption:** Promote the implementation of cloud-based inventory management by providing a right employee training and onboarding. Ensure your staff's performance is optimize with specially designed training initiatives that address your goals and objectives.

4.3 Enhancing Customer Experience with Data Insights

It has never been as significant as it is now to comprehend customers' wants, choices, and patterns as organizations proceed to advance in the digital world. By incorporating these tools into the organizational strategies, there will be better understanding of the customers thus a way of improving on customer satisfaction specifically customer loyalty(GENPACT, 2024).

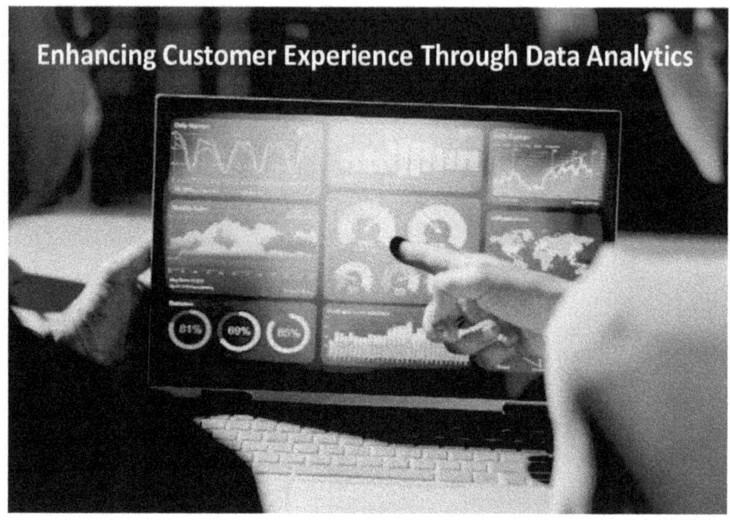

Figure 4.2: Enhancing customer experience through data analytics

Source: - *(Paci, 2024)*

Importance of Data-Driven Insights

They enable organisations to move from conventional customer categorisation to develop a better understanding of their customers. This may be accomplished by looking at customer data, such as age, gender, and spending patterns, among many other things. Businesses may learn a lot from this information, including how their customers behave while making purchases. These insights can then be used to create the customers experiences that would help in planning and delivering the customer experience strategies.(Solutions, 2023)

It is also possible to use data in order to define what part of the customers can be dissatisfied and how this can be changed. For instance, the feedback customers give can help managers determine the aspects that customers do not like about the firm's products or services for example the quality being offered. Using this information organizations are able to formulate means of enhancing these areas hence enhancing the level of customers' satisfaction. The information can also be applied to analyze the customers' behavior and promote customer loyalty programs. Customer information is important to businesses because it enables them to understand customers purchasing patterns, buying trends, and conduct. They will be able to see patterns of their customers and know what strategies they need to employ in order to re attract customers. For example, ecommerce merchants can by means of statistical models predict which of their customers are most likely to cease future patronage and formulate ways and means of retaining such customers.

Business intelligence also can enhance customers' experiences. The other way through which customer data can help businesses is in the recognition of issues that may hinder a customer or afflict the process by which he makes a purchase. It can also assist in building ways of how to overcome these aspects and enhance the comfort of the customer. This information can also be applied in marketing to enhance the marketing decisions so that the marketing return on investment could be boosted. Marketing communication can therefore be classified as communication that enables customer data to be collected, through which it can be discerned which marketing messages are appropriate and which are not. This can assist them in better understanding clients' needs and thus target their marketing strategies to only those, which promise the highest generating ROI.

It may also be of interest to some cable and satellite operators and media companies to create more SK-oriented customer experiences based on data. It is a fact that through customer data, firms and companies can get better understanding of the customers and how best they can create better solutions to address customer satisfaction and customer retention issues. When it comes to the customer, marketing focused analytics can help a business to understand what the customer wants and how best to offer them solutions to their needs thus improving the customer relations.

Steps to improving the customer experience with data and insights

1. **Let data be your guide**

   The deliberate application of pertinent propositional knowledge about the data and analytics is known as experience evolution. In order to provide your consumers a strategy, develop your brand, and improve your relationship with your audience, these insights will assist your firm in making changes to the goods and customer experiences you deliver. However, it may be both heroic and difficult to adopt change of experience as part of preparing your organisation for the experience evolution. Site-wide or app-wide data and analytics strategies are always the most effective, and they can even be applied to physical stores.

   When all your data are linked in a unified kind of reference library, you can determine the key KPIs, evaluate study results and apply those findings to the ideal delivery of superior customer value.

2. **Organize, collect, and store**

   In order to get ready for continuous examination of the customer's journey and experience, you must find and gather the relevant information.

   The first step is to ascertain if your business has a single digital channel for direct-to-consumer sales or whether you need to merge many databases. Next, learn what data is accessible and how it is gathered for each channel. Lastly, make sure that all data is safeguarded and that regulations are taken into account.

   This final stage is crucial, particularly since many organisations haven't taken data regulation into account.

   It's easier said than done to complete all of these tasks. Only 20% of businesses can easily handle data privacy compliance, according to eMarketer.

3.   Establish a data framework

A strong data collection is insufficient. To assess, learn from, and act upon data-driven insights, you must create a framework. In essence, the framework you create at the beginning serves as a guide for future data and analytics use.

In order to evaluate client behaviours and motives, it is important to avoid being overly fixated on KPIs and instead do more thorough study. You may use these insights to provide your clients with more value.

Last but not least, your capacity to assess and respond to data insights in real time will determine how quickly you can adapt to change.

4.   Reap the benefits of data-driven transformation

A strong data management is a crucial determinant of success for every organization regardless of its size, industry or sector.

Establishing a data-first strategy always starts with thinking about the outcome you hope to get. Next, consider how data may improve your relationship with your clients and enable your staff to make wise decisions.

Activating data insights throughout the company will let your company change and adapt to any circumstance. Therefore, concentrate on adjusting your strategy to support the success of your clients and staff. Additionally, this endeavour will assist you in proving the return on your investment to the larger company.

The process of developing an always-on analysis framework is not a one-time task that is limited to your analytics team. Delivering the best possible client experience, updating your digital channels, and reviewing your strategy are all necessary as you begin to understand and modify your KPIs.

4.4 Personalization Strategies Using Cloud Analytics

Courtesy has recently become one of the major sources where retailers differentiate their service offerings to fit the needs, wants, and demands of the market. Cloud analytics enable retailers to collect enormous amounts of data, process it and subsequently use insights derived from the data to market to customers across touchpoints. The collection of information from many other sources such as web interactions, past purchases, social media presence and demographic information compels cloud analytics to deliver accurate and real-time customer personalization techniques which in turn boost customer satisfaction, loyalty and profitability. (SG Analytics, 2024)

Key strategies of Personalization of services:

1. **Smart Product Suggestion**

    Customer behaviour is analyzed with the help of machine learning algorithms to define trends in the customers' purchase behaviour patterns on the cloud analytics platforms. From the above data, customized recommendations can be given by the retailers as they tap real-time information regarding customers. For instance:

    -   Online stores suggest "Customers who bought this also bought" products.

    -   It involves recommending specific products with the help of the list created according to the history of site visits or items added to the cart.

    -   Products that are seasonally advisable or following the current trend are associated with a particular demography; and improved chances of buying.

2. **Targeted Marketing Campaigns**

    Retailers can even target specific customers' behaviors and physical location, or any other characteristic that can be used in dividing customers into different segments, making it easier to develop very relevant marketing communications. Examples include:

- Direct mail or electronic marketing communications with an offer of products that the customer has some previous exposure to or related products on which may be a promotion.

- Specific advertisements of social media hitting specific segments of the population based on predictive analysis.

- In-store promotions, for instance, coupons sent by notification to a client who is probably near a retail store.

3. Customized Pricing and Offers

This way, cloud analytics help firms to develop and deliver appropriate redemptions of discounts, rewards, and promotions for loyalty or regained customers. Key applications include:

- Also, customer profiling for targeting with VIP discount rate or with a heads-up sale event.

- Giving retention incentives to customers who seem likely to defect, like flier programs for products that are commonly bought.

- There are also product bundles that customers are likely to use together or increase the quantity that will be able to be sold.

4. Integrated and Connected Cross-Channel Interactions

Cloud analytics automatically combines data from physical stores, e-commerce sites, mobile devices, and contact centers, giving each customer a single picture. This seamless integration supports:

- Enabling customer to see the products displayed physically in their physical stores or place an order with options of pickup in stores present.

- Alerting a customer about specific offers when they are close to a store.

- Helping customize the customer interaction within a call by arming a representative with information about the customer.

5. **Predictive Personalization**

Like a womb, cloud platforms predict customer needs and trends and can deliver personalized services. For example:

- When the preceding product is a consumable product that has to be replenished periodically, implementing a replenishment suggestion.

- Sending product offers before the onset of particular holiday/ season (for example, sending invitations to buy winter clothes before the holidays).

- Target marketing promotion in line with event derived from the buying process, like weddings purchasing of homes among others.

Advantages of Personalization with Cloud Analytics

1. **Enhanced Customer Engagement**

Customer engagement will be enhanced hence will enable the retailers to provide customers with what they need. For example, a whitepaper acquired in a more targeted and personalized fashion will be more likely to be opened and clicked on as opposed to a generic piece of mail.

2. **Improved Conversion Rates**

This is because the use of recommendations will enhance purchase likelihood among customers. Grocers and other store operators have achieved sizable gains in sales as they market goods by targeting consumers, avoiding the abandonment of purchases and encouraging purchases on the spur of the moment.

3. Greater Customer Loyalty

Customer satisfaction can be attributed to an organization demonstrating an understanding of the customer, and as Customers are therefore more inclined to stick with a specific brand. Traditional loyalty schemes combined with analytics that work in the cloud offer the right incentives to buy from the same outlet.

4. Competitive Advantage

Such retailers can thus create competitive advantage out of what can be seen as a commodity market through providing superior customer experiences by utilizing cloud analytics for personalization.

5. Data-Driven Decision Making

The information gathered from the cloud analytics in real time enables the business to dynamically adapt to change hence people are assured that the personalization methods being adopted is relevant.

Challenges and Considerations

Despite its benefits, personalization through cloud analytics comes with challenges:

- **Data Privacy and Compliance:** Modern acts such as GDPR and CCPA demand organizations and businesses to manage and process customer data appropriately. There are now responsibilities that retailers must meet in regard to data collection, and one of them is transparency along with choice.

- **Avoiding Over-Personalization:** When combining personalization at both of these levels, organizations may overdo it, and this is likely to compromise customer experience. At the same time, retailers have to find the ways to use personalization in moderation and not overpowering.

- **Integration Complexities:** Integrating information from multiple systems into a single application may be complex, and it may take a lot of money.

- **Scalability and Performance:** While MEC produces more customer data, he appreciates the need not to have cloud systems slow or strained.

Cloud analytics-based personalization tactics are making a profound impact on the way retail is being conducted by helping the business engage more closely with the customers. Through adaptive recommendations, constructing targeted advertising strategies, and providing uninterrupted customer service manner, retailers can user focus, brand fidelity, and sales. Regardless of the difficulties such as data privacy and system integration that need to be solved, the positive impact of applying cloud analytics for personalization has been suggested to overcome these obstacles. The mentioned strategies help firms to ensure sustainable development and strengthen their positions in the growing competition in the retail market.

4.5 Supply Chain Optimization via Cloud BI

In an era dominated by data, A vital tool for making wise judgements in a variety of businesses is business intelligence (BI). Bl leverages data collection, analysis, and presentation to improve business operations, making it a vital component in today's fast-paced market environments. (Sharma, 2024).

Regarding supply chain management, Bl tools have transformed traditional practices, enabling businesses to respond more swiftly and efficiently to market demands and logistical challenges.

Enterprise business intelligence is more than just a technological advancement that enterprise businesses need to adopt; it is a strategic asset they need to gain a competitive edge. As industries continue to evolve, the role of business intelligence for supply chains will expand beyond simple data management to become a core element of operational strategy and innovation

Supply chains usually involve various stages and stakeholders, from production to delivery, which can get complicated.

BI tools simplify this by providing clear and directly applicable insights, enhancing decision-making, and boosting operational efficiency.

By analyzing data trends and patterns, businesses can more precisely forecast future scenarios, improve planning, and adapt quickly to market shifts.

In addition, Bl has created a culture to proactively identify and mitigate on risks that exist in supply chain. Advanced analysis helps to predict future problems and to adjust the strategies providing the growth of the company. This ability is particularly useful in today's environment characterized by extensive and sensitive supply chain networks that originate from different parts of the world, and may be impacted by changing global political and environmental conditions.

In this blog, we will explore how Business Intelligence not only follows the changes that occur in supply chain management constantly but it actually initiates these changes. We will discuss specific Bl tools and methods at the forefront of transforming the industry, showing how they aid in cost reduction, operational efficiency, and service delivery improvement. So, without further ado, let's begin.

What is Supply Chain Business Intelligence?

Supply chain business intelligence is the strategic use of analytical tools and technologies to draw actionable insights for greater operational efficiency throughout the supply chain. It's the collection, integration, analysis, and combination of different data sources, from suppliers through manufacturers to distributors, retailers, and customers.

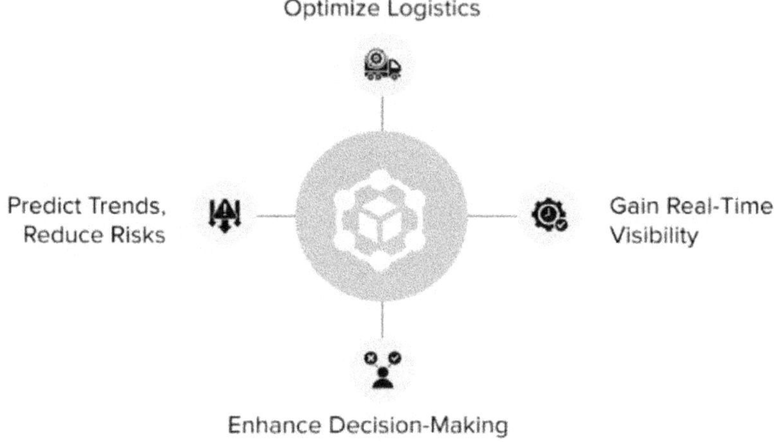

Figure 4.3: Transform your Supply Chain with Business Intelligence

Source: *- (Sharma, 2024)*

Benefits of Business Intelligence in Supply Chain

Business Intelligence The goal of supply chain management (SCM) is to improve the flow of its supply chain by using data methodologies and supply chain analysis. BI tools collect data from multiple supply chain sources, analyses it, and display it so that organizations can readily comprehend difficult situations. As a process, data analysis entails the transformation of raw data for use in making sound decisions in organizations with speed and precision.

A McKinsey report titled: "The Data-driven Enterprise of 2025" suggests that firms investing in Business Intelligence solutions are right in line for a dramatic makeover. Embedding big data into their strategic decision making and operations will improve decision making and organizational performance. All these enhancements are expected to afford yet other increments in supply chain management, sensitization and flexibility to market conditions.

Now, let us examine the resolutions of business intelligence in the supply chain as explained below:

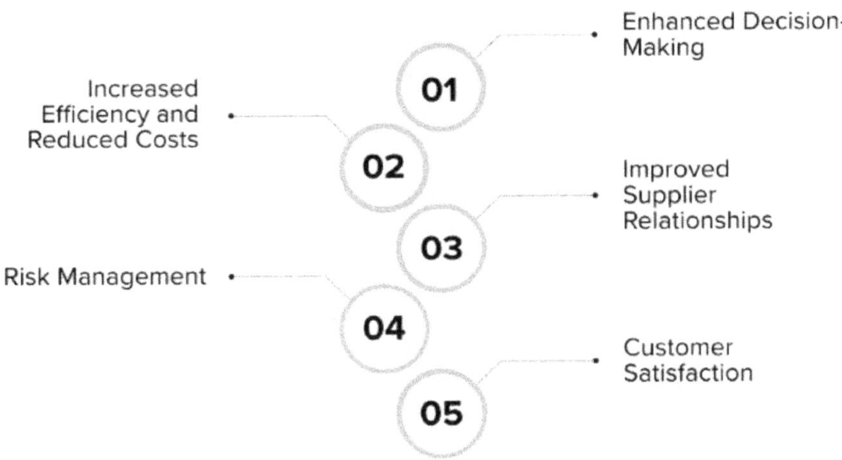

Figure 4.4: Benefits of leveraging BI solutions for your supply chain operations

Source: *- (Sharma, 2024)*

1.   Enhanced Decision-Making

Another key advantage of Bl in portfolio supply chain management is enhancement of decision making. Real-time information also enables managers to easily notice places that must be addressed and make wise decisions regarding it. For instance, it is possible to find that a certain product is not often available in stock. In that case, Bl can help develop a solution, it may be a supplier issue or a forecasting inaccuracy, that will encourage managers to act.

2. Increased Efficiency and Reduced Costs

The use of BI in the supply chain management is mainly to reduce daily effort as most processes involve analyzing data. This helps to minimize the number of people needed to keyed and review data thus lowering the cost of labor and minimizes errors that people possess. Further, BI may help to find the best way for delivering goods and transportation, which will consequently decreasing the fuel expenses and deliver the goods faster and increasing the customer's satisfaction.

3. Improved Supplier Relationships

Better relationships with the suppliers are one of those aspects, which can explain what business intelligence in the supply chain is. By implementing BI, companies are able to share appropriate data in relations to purchase pattern and inventory requirements so as to improve the business relationships of the two companies. This transparency results to; Reliable supply chain flow since the buyer's suppliers and any flow in between can adapt to any change easily and amicably.

4. Risk Management

Bl also strengthens a business risk management procedure by offering tools that can forecast supply chain break down. For example, if there is a business intelligence related to supply chain that shows that there might be a shortage of some raw materials sometime in the future, then the firm can approach different suppliers for other forms of materials to be supplied in future or look for other raw materials to use in the production . This proactive supply chain risk management approach assists the firm to prevent supply chain disruptions; thereby, sustaining production and consistent delivery schedules.

5. Customer Satisfaction

BI helps to enhance customer satisfaction as a crucial factor of every organization. BI tools can assist a company in determining, based on data and feedback derived from customers, which products are likely to be needed and when thereby improving a company's stock control and subsequently, reducing on out of stock merchandise. Additionally, improved delivery times and the ability to quickly resolve issues contribute to a better overall customer experience.

4.6 Addressing Data Security in Retail Cloud BI

Cloud data security is the process of protecting data and other digital information assets against security attacks, human error, and insider threats. In cloud-based settings, it uses technology, rules, and protocols to safeguard your data while granting access to those who need it.

The ability to access data from any device with an internet connection is one of the many benefits of cloud computing, which also improves scalability and agility and reduces the risk of data loss during catastrophes or outages. However, many organisations are still hesitant to migrate sensitive data to the cloud because they struggle to understand their security options and meet regulatory obligations (cloud, 2024)

As businesses move away from constructing and maintaining on-premises data centers, one of the most significant challenges they face is learning how to safeguard cloud data. What, then, is cloud data security? How is the security of your data maintained? And what best practices for cloud data security should you adhere to in order to guarantee the safety and security of cloud-based data assets?

Figure 4.5: Data Security in Cloud Computing

Source: - *(Patadiya, 2024)*

Why companies need cloud security

Businesses generate, collect, and store massive amounts of data on a continuous basis; this is the big data era in which we find ourselves. From highly confidential company or customer information to less delicate marketing and behavioral analytics data, this data covers a wide spectrum of uses.

Businesses are increasingly turning to cloud services to help manage the ever-increasing volumes of data, facilitate remote or hybrid workforces, boost agility, and decrease time to market.

As the traditional network barrier is quickly becoming obsolete, security teams are coming to the realization that their existing and past approaches to safeguarding cloud data require reevaluation. Since data and apps are no longer stored in data centers and more people are working remotely, companies must find ways to protect data and manage who has access to it as it moves across different environments.

Data privacy, integrity, and accessibility

Cloud data security best practices adhere to the same data governance and information security tenets:

- **Data confidentiality:** Only authorised individuals or procedures have the ability to view or alter data. To put it another way, you must guarantee that the information of your company is kept confidential.

- **Data integrity:** Data is dependable, accurate, and authentic, which makes it trustworthy. Implementing regulations or safeguards that stop your data from being altered or erased is crucial in this situation.

- **Data availability:** Data must still be available and accessible to authorised individuals and processes when needed, even if you wish to prevent unauthorized access. You'll have to maintain systems, networks, and devices operating efficiently and guarantee constant uptime.

These three guiding principles, which are often referred to as the CIA triad, are the bedrock principles of every effective security program or infrastructure. Any security incident, vulnerability, or assault will likely violate one or more of these principles. Because of this, security professionals use this process to evaluate potential threats to a company's data assets.

The challenges of cloud data security?

As more apps and data move away from traditional security mechanisms and architecture and out of a central data centre, the risk of vulnerability grows. While many of the same essential elements of data security on-premises remain, they must be adjusted to work in the cloud.

Common challenges with data protection in cloud or hybrid environments include:

- **Lack of visibility.** Companies have no idea what data and apps are housed where or what assets are in their inventory.

- **Less control.** They have less control over data access and sharing since third-party infrastructure hosts apps and data.

- **Confusion over shared responsibility.** If responsibilities and tasks are not clearly defined, there could be coverage gaps in cloud security, which is a shared responsibility between businesses and cloud providers.

- **Inconsistent coverage.** Numerous companies are discovering that multi-cloud and hybrid cloud solutions better meet their demands, but various providers have differing capabilities and coverage levels, which might result in uneven protection.

- **Growing cybersecurity threats.** As businesses continue to educate themselves on cloud data handling and management, internet criminals seeking a large payout find cloud databases and cloud data storage to be prime targets.

- **Strict compliance requirements.** Businesses are under pressure to adhere to strict privacy and data protection laws, which call for implementing robust data governance and enforcing security standards across various contexts.

- **Distributed data storage.** Lower latency and greater flexibility can be obtained by storing data on foreign servers. However, it can also bring up concerns of data sovereignty that might not be an issue if you were using your data centre.

What are the benefits of cloud data security?

1. **Greater visibility:** Your cloud's internal operations, such as the data assets you have, their locations, service users, and the data they access, can be monitored using robust data security measures.

2. **Easy backups and recovery:** By utilising the features and tools offered by cloud data security, your teams will no longer have to manually monitor and troubleshoot backups; instead, they can automate and standardise the process. Fast data and app restoration is another benefit of cloud-based disaster recovery.

3. **Cloud data compliance:** Data storage locations, access points, processing techniques, and security mechanisms are all part of a comprehensive cloud data security plan that is developed to meet compliance standards. Data loss prevention (DLP) in the cloud makes it easy to locate, sort, and de-identify sensitive information, which in turn reduces the likelihood of breaches.

4. **Data encryption:** Wherever sensitive data goes, it must be protected. When you work with a cloud provider, they will help you ensure the security of your data throughout transmission, storage, and sharing by using many levels of state-of-the-art encryption.

5. **Lower costs:** Cloud data security lowers the administrative and managerial load as well as the total cost of ownership (TCO). Furthermore, with automation, simplified integration, and constant alerting, cloud providers make it simpler for security professionals to do their duties by providing the newest security features and capabilities.

6. **Advanced incident detection and response:** One advantage of cloud data security is the availability of modern AI technologies and integrated security analytics. These allow for the automatic detection and resolution of security incidents, as well as the automatic search for suspicious activity.

4.7 Case Study: Retail Chain's Transition to Cloud BI

The management of a global retail chain operating hundreds of stores across the world experiences huge problems in inventory control, improving customer experience, and designing an efficient supply network. It had legacy systems that could not handle large chunks of data generated by store and online business; it experienced issues such as stockouts, overstocking, and lost opportunities for interacting with customers. To solve these problems the retail chain switched to Cloud Business Intelligence (Cloud BI) which completely changed its functioning (Olexová, 2014)it analyses the Business Intelligence life cycle; it evaluates factors impacting the adoption from the Diffusion of Innovations perspective. One of the findings is that requirements engineering is critical, and even small issues have a tendency to cause big problems. This links to the sentiment among managers, often worrying that IT projects will run over-budget and/or over-time. Finally, the presented research identifies benefits that are considered to be the most important by the retail chain managers. An important finding is that managers consider improved decision-making to be the most significant benefit.","author":[{"dropping-particle":"","family":"Olexová","given":"Cecília","non-dropping-particle":"","parse-names":false,"suffix":""}],"container-title":"WSEAS Transactions on Business and Economics","id":"ITEM-1","issue":"1","issued":{"date-parts":[["2014"]]},"page":"95-106","title":"Business intelligence adoption:Acasestudyintheretailchain","type":"article-journal","volume":"11"},"uris":["http://www.mendeley.com/documents/?uuid=cf60b247-0f0f-4ea1-b227-9774cff655fb","http://www.mendeley.com/documents/?uuid=04121e52-0e90-4a7f-8a35-6e3f77759caa","http://www.mendeley.com/documents/?uuid=15540183-8496-465d-8b93-ab2ea439910b"]}],"mendeley":{"formattedCitation":"(Olexová, 2014.

Implementation of Cloud BI

This retail chain with a Cloud BI provider to integrate the business data into a unified, more extensible cloud environment. This integration enabled real-time analytics and provided actionable insights across multiple functions:

- **Inventory Optimization:** The chain used Cloud BI in the determination of the sales and the probable demand so as to be able to change the stocks physically in a choreographed manner. Drawing on predictive analytics, it cut down the incidence of overstock to the barest minimum and also managed to reduce incidences of stockouts, thus making sure that products were in the right places at the right time.

- **Personalized Customer Experiences:** The solution enabled the chain to develop additional deep customer insight by linking purchase patterns, Web activity, and demographics. Sophisticated calculations yielded product recommendations and unique deals that enhance the buyers' experiences and thus retain consumers.

- **Supply Chain Efficiency:** Supply chain reporting ensures the chain captured real-time data that helped to define areas of slow movement or slow logistics. The suppliers and the distribution centers worked much closely together with the research showing that the delivery time incorporated far less operating expenses.

- **Data Security:** For solving data security issues, the chain deployed high encryption levels, access rights within the Cloud BI platform and monitoring tools. This helped in following the industry retail data protection laws in avoiding loss of customer and its operation data.

Results and Impact

The transition to Cloud BI delivered transformative results for the retail chain:

1. **Increased Sales:** Its effects included better cross selling and upselling by 20% and this was achieved through personalized recommendation options and target marketing.

2. **Operational Efficiency:** You can increase real-time inventory data availability and decrease overstock by 15% helping to increase stock availability and customer satisfaction.

3. **Enhanced Supply Chain Performance:** Logistic costs were cut by a tenth which enabled efficient delivery time of products by an additional ten percent and overall operation cost was lowered by twelve percent.

4. **Data-Driven Decision-Making:** Real-time analytics and trends with predictive analysis helped managers by having timely data-based decision making which increased business flexibility.

4.8 Chapter Summary

This chapter analyses how Cloud Business Intelligence (Cloud BI) is transforming the retail industry through the enhancement of business-critical processes as well as by providing superior customer experiences. Cloud Inventory Management makes it easier for businesses to have control of products or services inventory through real-time updates on the products, improved supply chain processes and improved processes of updating the inventories.

Insights developed from cloud analytics help retailers to drive smarter and more effective communications. Measures like real-time offers and promotions, one-to-one communications, next-best action, and flawless cross-channel conversations enhance customer retention and increase consumption. The combination of machine learning and real-time data analysis caters to Marketing campaigns and business operations generating a competitive advantage. In the context of Supply Chain Management, BI tools support a more rational approach to decision making regarding visibility, forecasting and risk. These tools enable the retailers to manage their stock levels, minimize operating expenses and costs of dealing with suppliers while still meeting the needs of their customers.

To protect consumers' and business data, cloud data security is important for organizations. They consist of compliance with data privacy statutes, data encryption, data confidentiality, data integrity, and data availability. While it has certain drawbacks, for instance, visibility and confusion over who is responsible for security, cloud security provides opportunities for automated backups, extremely accurate threat identification and cost efficiency. An example from an international chain of stores shows how Cloud BI can change the business. What specific performance improvement can be claimed by the retailer after the implementation of the integrated data environment? The cross-selling and upselling have been increased by 20%, the operating efficiency has increased by 15%, and logistics costs/delivery time has decreased by 10%. Overall, Cloud BI acts as a disruptor, innovating the retail industry, by driving efficiency, improving decisions, personalization and security, which, therefore, guarantees continuous growth within a growing market.

Multiple Choice Questions (MCQs)

1. **What is the primary focus of the chapter on Cloud BI applications in retail?**

    a. Marketing strategies in retail

    b. Optimizing inventory management and customer experience

    c. Developing mobile applications for retail

    d. Improving retail store layouts

2. **How does Cloud BI optimize inventory management in retail?**

    a. By offering real-time tracking of stock levels

    b. By automating customer feedback

    c. By increasing the speed of store openings

    d. By reducing employee workload through automation

3. **Which of the following is a key benefit of Cloud BI in enhancing customer experience?**

    a. Reducing employee turnover

    b. Providing data-driven insights into customer preferences

    c. Automating store operations

    d. Limiting customer interaction

4. How can retailers use Cloud Analytics for personalization?

a. By collecting generic customer data

b. By analyzing transaction data to recommend personalized products

c. By reducing the variety of products offered

d. By offering discounts to all customers

5. What is one-way Cloud BI aids in supply chain optimization?

a. By eliminating the need for physical warehouses

b. By improving visibility into supplier performance and demand forecasting

c. By removing middlemen in the supply chain

d. By limiting the number of suppliers

6. What is a major concern when using Cloud BI in retail?

a. High initial setup costs

b. Data security and privacy

c. Reduced customer engagement

d. Inconsistent inventory data

7. What is the focus of the case study in the chapter?

a. The impact of social media on retail sales

b. A retail chain's transition to Cloud BI for improved decision-making

c. The history of Cloud BI applications

d. The challenges of brick-and-mortar retail stores

8. **What does Cloud BI allow retailers to do with their customer data?**

    a. Store it without analyzing it

    b. Collect it from a single channel only

    c. Analyze it to understand customer behaviors and preferences

    d. Share it freely without restrictions

9. **Which aspect of Cloud BI addresses potential risks in retail data management?**

    a. Personalization of customer offers

    b. Ensuring compliance with data security regulations

    c. Increasing the number of data storage options

    d. Automating customer service responses

10. **What is a significant advantage of using Cloud BI for retail businesses?**

    a. It reduces the need for customer interaction

    b. It provides scalability and flexibility in data processing

    c. It limits access to real-time data

    d. It only works for large retail chains

Answer

1	2	3	4	5	6	7	8	9	10
b	a	b	b	b	b	b	c	b	b

Chapter 05

MANUFACTURING TRANSFORMATION WITH CLOUD BI

5.1 Chapter Introduction

In the evolving age of manufacturing organizations adopting Cloud Business Intelligence or Cloud BI, it has become crucial for change and innovation. This chapter focuses on how Cloud BI helps manufacturers improve production, apply predictive maintenance, and integrate the data from the IoT for better operations. It shows how new functions utilize advanced analytics for better quality control and which problems may arise during cloud implementation. The chapter also illustrates through examples a variety of real-life companies that show how manufacturers can use Cloud BI to attain greater levels of intelligence in production and remain relevant to a dynamically changing environment.

5.2 Improving Production Efficiency through Data Analysis

Companies now gather, retain, and utilize more data than they did in the past. The amount of data being generated daily is extraordinary and only going to increase, in addition to the fact that there is more data available. Businesses may fundamentally transform their operations by using this data. Particularly potent is data analytics in industries like manufacturing and others that mostly depend on procedures and technology.

Although industry has always employed data, the capabilities of data analysis today are far more advanced than those of even the recent past. Therefore, in order to stay competitive, factories must implement modern data analysis techniques that identify trends, areas for improvement, and little inefficiencies that are aggravating (impact, 2024).

The Role of Data Analytics in Manufacturing

Data analytics is essential to contemporary manufacturing because it turns unprocessed data into insights that can be used to improve production, efficiency, and decision-making. Manufacturers are able to forecast results, find trends, and streamline operations by gathering and evaluating data from different manufacturing phases(Tiwari, 2024).

This data-driven strategy enables real-time monitoring and control, which significantly raises quality, lowers costs, and decreases downtime.

In manufacturing, predictive maintenance is one of the principals uses of data analytics. The ability to predict when a certain piece of equipment will develop a problem through the monitoring of sensor data and data from several pieces of machinery allows one to arrange maintenance time to prevent needless production loss. It not only saves production from stopping, but it also helps to extend the life of any equipment. Additionally, analytics increase quality control by identifying defects or noncompliance with specifications, reducing material consumption, and enhancing product quality(Potter & Broklyn, 2024).

In the area of SCM, data analytics offer good returns in demand forecasting, inventory and stock control and even, logistics. To enhance supply chain efficiency, manufacturers also have to determine how best to deliver the necessary supplies to produce their products, and to deliver their goods to the market as efficiently as possible, while avoiding excessive investments in inventory.

In conclusion, data analytics in manufacturing enhances the organization's operation by increasing profitability, efficiency and industry competitiveness and provides support for strategic decision making and market conditions.

Some of the other key benefits of data analytics in manufacturing include:

1. **Predictive Maintenance:** Expect equipment breakdowns and plan for their consistent repair to help prevent to reduce the odds of pricey downtimes and improve machinery durability.

2. **Quality Control:** Prevent the construction of defective products and identify the inefficiencies that cause unnecessary waste and nonconforming product in the course of the construction process.

3. **Process Optimization:** To analyze possible weaknesses within the processes to maximize and optimize the improving in productivity.

4. **Supply Chain Efficiency:** Make inventory management, demand forecasting for product and other relevant to be on time and cut costs of logistics.

5. **Cost Savings:** Sustained production cost through better economy, lower scrap rate and better utilization of resources.

6. **Enhanced Decision-Making:** This will need great management of the available data in order to offer the best strategies and competitive edge in the market.

7. **Innovation and Competitiveness:** Promote innovation through the use of insight to enhance product as well as enhance performance levels within commercial trends.

Types of Data Analysis in Manufacturing

The processes of manufacturing may well be intricate and highly technical to a point that if manufacturing workflows are compared to other industrial processes, there exist many subcategories of manufacturing that would require deeper analysis of data.

Descriptive analytics, for instance, summarises historical trends in order to comprehend how well the company performed in the past. This type of analysis creates a summary of the events that have transpired over a certain amount of time by using dashboards and charts that provide a graphical depiction of what has happened. Setting up the starting parameters and examining a certain production period are two applications for this type of study.

Predictive analytics uses statistical and machine learning approaches to forecast future states based on historical data, whereas decision support simulates and idealizes the decision-making process. It involves predicting supply chain interruptions, demand fluctuations, and equipment failures in the industrial sector. Many risks may be reduced by these occurrences, and the machinery's maintenance plan can be adjusted to meet the anticipated demand.

The standards are, in any event, intended to improve the efficiency of resource management and the flow of activities.

Big data may be used in manufacturing to optimize supply chains, better allocate resources, plan production as effectively as possible, and make better decisions based on the data gathered. Since there will be a better understanding of the hole manufacturing system, its effectiveness, and its competitiveness in the global manufacturing market, the use of various analytical approaches in manufacturing will eventually greatly aid in decision-making within the manufacturing sector (impact, 2024).

5.3 Implementing Predictive Maintenance Strategies

Let's take a quick look at what predictive maintenance is before we go into the solutions. Predictive maintenance employs a proactive strategy by evaluating real-time data to forecast when equipment failure is likely to occur, in contrast to traditional preventive maintenance, which depends on planned inspections and maintenance tasks. Predictive maintenance algorithms can detect trends and anomalies that point to possible problems by tracking and evaluating a variety of data, including temperature, vibration, and performance indicators.

There are several advantages to predictive maintenance. It increases equipment availability, minimizes the chance of unplanned malfunctions, prolongs the life of assets, lowers maintenance costs by preventing needless repairs, and maximizes spare component inventories (sensemore, 2024).

- **Strategy 1: Establishing Clear Objectives**

  Having well-defined goals is the first step to any fruitful endeavour. Clearly defining and aligning your goals with your overall business objectives is of utmost importance when implementing predictive maintenance. Determine your goals for predictive maintenance. In other words, is the goal to facilitate organisational efficacy, raise the possibility of equipment success, or decrease operating expenses? Lastly, it is easy to state strategies to get there and identify how to tell that one has arrived provided objectives are clearly defined.

  Equally crucial is being aware of which Key Performance Indicators (KPIs)—both broad and narrow—are relevant to your objectives. Some examples of key performance indicators include asset availability, maintenance expenses, mean time to repair (MTTR), and mean time between failures (MTBF). To evaluate the efficacy of a predictive maintenance program and make course corrections as necessary, one can use the following KPIs.

- **Strategy 2: Data Collection and Integration**

  Information is the blood solution of predictive maintenance. In order to optimize maintenance, there is a requirement to gather and collect specific data and information for operation of predictive maintenance applications. It can be sensed data, IoT data, logs from maintenance and repair work, historical data, and much more. The area is one where the focus is to collect large sets of consistent and reliable information that reflects the status and productivity of an asset.

This way it is easier to get the whole picture of your assets and, relying on analytics, find patterns or links that might not be noticeable otherwise. For example, if data coming from the vibration sensor, temperature sensor, and other maintenance records are analyzed collectively, it may point out specific regular and irregularities that may really help in getting accurate equipment failures. The last trend to consider is making sure that data collection and integration is fully automated to allow for quicker results without a chance of human mistakes slowing things down.

- **Strategy 3: Data Analysis and Machine Learning**

  However, after data collection and data integration activities, the subsequent step is the effective analysis of the data. Architects and designers can, however, use advanced analytics and Machine Learning algorithms in creating a new systematic cognition. This is because by using statistical approaches and other artificial intelligence technologies to analyze the data you are able to see patterns, relationships, and even data outliers pointing to probable equipment breakdowns.

  Iteratively processed data sets can yield better results for machine learning algorithms since they are data-driven. It has the ability to spot even the smallest changes in asset behaviour and notify maintenance personnel so they can take the appropriate action. An algorithm may determine, for example, that a machine's vibration level progressively increased and indicated bearing failure based on its analysis. By using this knowledge, maintenance staff may do repairs when the failure hasn't occurred or hasn't occurred seriously, which lowers the potential expenses from failure.

- **Strategy 4: Condition Monitoring and Sensor Technologies**

  To successfully develop and execute a solid pathology of predictive maintenance, the plan requires buying in condition-monitoring methods and sensor devices. Condition monitoring is much like assessing the status of the assets in the manner that checks for any form of variation from the standard running conditions. This can be achieved by using various technologies of sensors according to the real data they collect.

  For instance, the sensors of vibration can help flag mechanical problems with equipment simply by noting shifts in vibration. Temperature sensors can give indications of changes in temperatures that might signify that an engine is overheating or if cooling systems have failed. Pressure sensors, oil analysis sensors, or acoustic emission sensors that are of a different type to vibration measurement may also be useful for gaining status on an asset.

  Condition monitoring has become even more powerful with the invention of the Internet of Things (IoT). These sensors have the capacity to pick real-time data and forward it to a central system for processing. This makes it possible for maintenance teams' products and assets to be supervised from other quarters besides receiving alerts when something amiss is recorded (sensemore, 2024).

- **Strategy 5: Implementing Predictive Maintenance Software**

  In order to correctly facilitate and monitor the overall process of predictive maintenance, it is necessary to acquire the proper tools, in this case – predictive maintenance software. These software solutions offer functions for data management, processing and representation which afford the overview of asset state and the schedule of maintenance operations.

  Selecting the right predictive maintenance software for your business involves factors including; scalability, integration, user-friendliness and intelligence. This is where one should look for software solutions that include analytics for business, customizable dashboards and auto reports. An effective software solution for your predictive maintenance will create a huge boost in your maintenance processes and decision-making.

- **Strategy 6: Building a Skilled Workforce**

  The workforce has to be skilled for it to be more capable of performing predictive maintenance if required in an organization. In the case of predictive maintenance, it will be prudent to spend good money on training and skill development of your maintenance staff and personnel.

  Summarize how key competencies for predictive maintenance calculations, analytics, and knowledge of sensors. A number of training programs and workshops should be implemented for your employees in order to have those necessary skills. Promoting a culture in which people are recommended to improve and educate and learn more about the new technologies and methods for predictive maintenance.

- **Strategy 7: Establishing Maintenance Workflows**

  Define the maintenance work flow in a way that facilitates the successful application of predictive maintenance techniques. This may be accomplished by redesigned dependability monitoring techniques that consider the assets' criticality, data analysis findings, and skill levels. This will help you choose where to allocate the majority of your maintenance efforts and resources.

  A description of some of the expected responsibilities of your maintenance crew should be provided. Provide specialised staff to handle data gathering, analysis, and upkeep tasks. Create feedback loops to gather information from maintenance tasks and apply it to processes for further maintenance. In order to enhance your predictive maintenance plan over time, you must constantly improve.

5.4 Integrating IoT Data with Cloud BI Solutions

Have you ever wondered how much-unrealized potential your company's everyday activities hold? Imagine being able to use every bit of information from every digital interaction, sensor, and gadget. The Internet of Things (IoT) is more than just a catchphrase; it is a revolution in data collection and use that is now taking place. However, how can this enormous ocean of data be turned into insights that can be put to use? (medium, 2024).

Figure 5.1: IoT Data with Cloud BI Solutions

Source: - *(medium, 2024)*

It is anticipated that over 25 billion IoT devices will connect and transmit data globally during the next six years. The worst part is that without the proper instruments to combine and evaluate this data, it may as well be invisible. Business analytics tools can help make the invisible essential in this situation.

Why Use Business Analytics Software with IoT Data Integration? Consider the last important business choice you made. Did it follow statistics or was it based on a gut feeling? Adding IoT data to your business analytics dashboard does more than simply increase the volume of your data; it changes your strategy from being reactive to being proactive and from relying on guessing to being precise.

Key Technologies Facilitating IoT Data Integration

- **APIs (Application Programming Interfaces):** The key to integrating IoT data with corporate analytics tools is APIs. Without the need for human involvement, they provide smooth communication between various devices and systems. Businesses may make sure that data gathered from different IoT devices is consistently fed into business analytics systems by using APIs. This improves data timeliness and dependability, which are essential for successful BI analytics, in addition to streamlining procedures(Chatterjee, 2024).

- **Middleware Solutions:** In order to convert and transport data from IoT devices to analytical tools, middleware serves as a crucial middleman. In order for corporate analytics tools to more effectively process data from many sources, this technology coordinates data formats, protocols, and communication. Middleware solutions' large data processing capabilities imply they can meet the scalability needs of comprehensive IoT frameworks. To further facilitate decision-making using business analytics, they also strive on data transformation to guarantee that data collected from IoT devices is in the optimal format for analysis (Ngu et al., 2017).

The Role of Cloud Computing in IoT Data Management

- **Scalability and Flexibility:** Through cloud computing, it's easy to support the huge data received from the IoT devices due to the scalability it offers. The newer generation of IoT deployments is accommodated nicely because cloud platforms are able to increase the amount of resources allocated to a given app to handle the data load without a decline in performance. This elasticity is important for organizations that process large/unstructured data sporadically to guarantee that their business analytics dashboard will be stable and dynamic.

- **Data Integration and Accessibility:** Cloud services have the potential to serve as a central site for gathering, analyzing, and storing all IoT data. Because of this centralization, almost any data from any other source may be combined into a single integrated business analytics site. Furthermore, cloud computing has made real-time data accessible, which is essential for sectors that depend on it to make choices instantly (medium, 2024).

The Impact of Edge Computing on IoT Integration

- **Real-time Data Processing:** Edge computing enables calculations to take place near or at the edge of the point of creation (i.e., Internet of Things devices). Because it facilitates quick decision-making, this closeness lowers latency in BI analytics solutions. Certain procedures call for a reaction in a brief amount of time; even a single second might result in a significant interruption to corporate operations or a missed opportunity.

- **Bandwidth Optimization:** As per Edge computing techniques, most of the data analysis results in not sending data back to central servers or cloud systems. It was also found to eliminate bandwidth utilization to minimal levels while also cutting across the costs of transmission and congestion. From this perspective, it means optimization of business processes and cutting of costs by businesses, therefore explaining why business analytics is crucial in determining the right resource utilization.

Step-by-Step Guide to Integration

Step 1: Evaluate Your Current Infrastructure

First, assess how well-prepared an organizations' IT is to handle the strain that IoT data is putting on it. This involves evaluating your computer's processor power, storage options, and networking aspects.

Additionally, you should pinpoint the precise locations of your company's various IoT sensors and equipment. Knowing what kind of data each device generates and if it relates to the objectives of business analytics is also crucial.

- **Infrastructure Optimization:** Besides analyzing, think about how to improve your current resource base. Introduce techniques in edge computing in order to enhance data processing an IoT devices in order to reduce latency and use less bandwidth. It also helps to maintain a structure that improves your infrastructure in the management of real-time data necessary when a BI system is growing evanescent.

- **Integration Capacity Planning:** Develop a model that correctly estimates future demands for infrastructure based on assumptions regarding the growth of IoT information. It assists in making the overall systems sustainable and responsive or, in other words, elasticity to meet future demands.

Step 2: Set Clear Integration Goals

Define what you expect to get from your IoT integration. Regardless of whether it is making operations more efficient, making customer experiences better, or increasing prediction in terms of equipment maintenance, aim and objectives will determine the integration process.

- **SMART Goals:** Make that your integration objectives are Time-bound, Relevant, Specific, Measurable, and Achievable (SMART). The stakeholders benefit from these accuracies in terms of progress tracking and establishing rationale in terms of ROI.

- **Feedback Loop Creation:** In order to regularly examine these objectives in accordance with the actual usage and situations of the software development process, it would be beneficial to keep up continuous communication with the technical teams and end users. Because of this, your initiatives are pertinent to both corporate needs and technological advancements.

It is particularly crucial for this reason that you be able to defend the investments and the impact of your IoT operations on the dashboard of business analytics.

Step 3: Choose the Right Business Analytics Software

Select a business analytics solution that works with the company's IoT data sources. The program should be scalable to manage extremely big data transactions and capable of operating in "real time".

- **Feature-Specific Evaluation:** Read more about additional products that specifically facilitate IoT connections, such as real-time insights from live data processing, predictions and warnings APIs that satisfy particular business needs, and analytical models that are appropriate for their intended use.

- **Vendor Collaboration:** Discuss potential service requirements and product modifications with software providers. The relevance of business analytics stems from the fact that procurement and sourcing, when combined with a strong vendor relationship, allow for the development of solutions that are appropriate for an organization's needs.

- **Customization and Scalability:** The software you choose should be able to adapt to your business needs and potentially offer solutions as your IoT usage grows. For this reason, a crucial solution—the instruments required to satisfy your company requirements—is business analytics software.

Step 4: Develop a Data Integration Plan

Recognize how IoT devices provide data to your framework. This includes identifying the data endpoints, the storage system, and any intermediary applications, such as middleware or APIs.

- **Advanced-Data Routing:** Create intricate data forwarding systems that can automatically route streams of IoT data to tools for further analysis based on the sources, types, and destinations of the data. This facilitates data management and speeds up results generation.

- **Compliance and Security Frameworks:** Provide complex security measures that adhere to international data protection regulations, such as the CCPA for Californian data and the GDPR for data from the European Union. In addition to protecting your data flows, this will assist stakeholders and company owners regain their trust.

To put it another way, creating a free alternative model of the final data usage and protection policy—which frequently specifies who may access the data—is necessary to achieve and preserve data purity.

Step 5: Implement Data Processing Mechanisms

Incorporate real-time processing tools to analyze data in real-time, as IoT data streams are typically near real-time. This is particularly crucial for applications that require immediate results, such as manufacturing process fault detection.

- **Data Enrichment:** To make the IoT data amount a more useful asset, supplement it with data from other business sources. This might broaden your business analytics dashboard by providing more details on the internal operations of the organisation.

- **Hybrid Processing Models:** It will be more efficient to combine the ideas of real-time processing for current data with batch processing for previous data. This two-pronged strategy allows you to get the most out of your data and use it for a variety of commercial applications.

- **Enhanced Data Enrichment Techniques:** Construct predictive analytics and anomaly detection by integrating IoT data with advanced AI algorithms. Your business analytics dashboard may become the king maker for your company using this, since it can help transform raw data into strategic knowledge.

Step 6: Visualization and Utilization

Provide your business analytics dashboard with your IoT data, and configure it to display the metrics that have been determined to be most pertinent to your integration. To help your staff understand how to use the new tools and data you have, you should conduct some workshops.

Adoption is necessary to effectively unlock and support the benefits that BI Analytics upgrades offer.

- **Interactive Dashboards:** Provide a basic user interface (UI) that allows the data from IoT devices to be dug down to the smallest detail. Allow users to customize the dashboard's interfaces for particular roles and orientations to increase its value and usefulness for a variety of users.

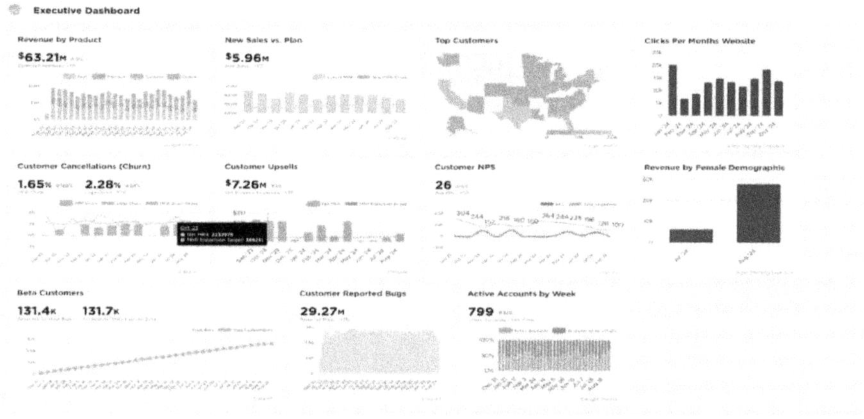

Figure 5.2: Interactive Dashboards

Source: - *(medium, 2024)*

- **Decision Automation:** Engage essential apps that carry out operations automatically based on certain KPIs shown in the dashboard. For instance, these applications can change supply chain parameters in real time or notify maintenance teams of potential equipment failures. By converting findings into simple processes, this automation improves the effect of your BI insights on your company.

Step 7: Continuously Monitor and Optimize

It is advised that you monitor the performance of your IoT integration to determine the extent to which it meets the established goals. This includes concerns about the quality, dependability, and timeliness of data collecting generated insights.

- **AI-Driven Optimization:** Utilize AI algorithms to continuously monitor and adopt efficient measures for the optimal integration implementation. In real time, this might highlight previously overlooked gaps and areas for development.

- **Sustainability Practices:** Include environmental management in your IoT monitoring and make sure to employ eco-friendly techniques, such as minimizing data transmission distance and appropriately managing storage systems. Additionally, it improves the concept of corporate social responsibility and maximizes system performance.

Since business analytics and the Internet of Things are developing domains, you must always seek methods to improve data integration and analysis in light of user input and emerging technology(medium, 2024).

5.5 Quality Control Enhancements via Cloud Analytics

There is lots of evidence that shows manufacturers are actually losing up to 40 percent of their operations to the quality costs. Even thriving companies lose 10-15%, underscoring the urgent need for effective quality improvement programs.

The absence of robust analytics capabilities represents more than a missed opportunity—it signifies potential revenue left untapped. Manufacturers, especially in high-tech sectors, are discovering that without comprehensive visibility into their operations, they risk leaving significant financial gains on the table. This realization has propelled the adoption of advanced analytics tools, both at the edge and in the cloud, to unprecedented importance.

Particularly, by leveraging real-time data from customer-installed products, high-tech manufacturers can proactively offer tailored maintenance contracts, ensuring equipment operates at peak efficiency. This not only generates new revenue streams but also minimizes warranty-related expenses by preemptively addressing potential issues. Such strategic use of analytics not only optimizes operational performance but also enhances customer satisfaction—a critical factor in maintaining a competitive edge in today's market (aceinfoway, 2024).

Key Quality Control Hurdles That Manufacturers Face Today!

Maintaining high standards of quality control presents several challenges for manufacturers that require careful management and innovative solutions:

- **Inconsistent Product Quality** – Real-time monitoring to detect and address variability in production.

- **Supply Chain Disruptions** – Enhanced visibility and predictive analytics to manage and mitigate disruptions.

- **Machinery Downtime** – Predictive maintenance to avoid the failures that are not expected yet.

- **Manual Data Entry Errors** – Reducing ambiguities due to use of Automated data collection and analysis.

- **Delayed Defect Detection** – Detection of defects at an early stage by undertaking periodic as well as systems analysis and review.

- **Limited Process Optimization** – Cognitive data analysis to reveal concerns and increase efficiency of manufacture procedures.

- **Regulatory Compliance Issues** – The tracking of compliance and documentation to make sure the company is meeting all industries standards.

- **Inadequate Root Cause Analysis** – Precise measurements as a tool to identify the main sources of quality problems.

- **Variable Supplier Quality** – Suppliers performance and quality issues have to be tracked on a continuous basis.

- **High Scrap and Rework Rates** – Conformance quality assurance in order to minimize scrap and rectification time.

- **Inefficient Inventory Management** – Optimized reliability and efficiency by using the data acquired to refine the forecasting of inventory demands.

- **Poor Production Planning** – Proactive planning and forecasting due to the capability of the system to forecast the future.

- **Lack of Real-Time Visibility** – Accurate, timely and encompassing understanding of all the processes happening in manufacturing.

- **Suboptimal Resource Utilization** – Optimal allocation of resources with the help of data gathered with the help to implement the process.

- **Difficulty in Scaling Operations** – Scalable cloud solutions to support growth and adaptation to market changes.

By leveraging predictive analytics, manufacturers can forecast potential quality issues, optimize production processes, and ensure compliance with regulatory standards. Additionally, cloud-based solutions facilitate departmental data integration, offering a comprehensive perspective of quality measures and facilitating data-driven decision-making.

Now as we move into 2024, the manufacturing industry is gearing up for significant technological advancements. According to Gartner, a staggering 54% of manufacturers are planning to increase their technology spending this year. This increase in investment demonstrates how important cutting-edge technologies like artificial intelligence (AI), the Internet of Things (IoT), and business intelligence are for promoting productivity, creativity, and competitiveness (aceinfoway, 2024).

5.6 Overcoming Challenges in Manufacturing Cloud BI

In an ever-changing business environment, manufacturing firms are increasingly using the cloud to boost operational effectiveness, cut expenses, and maintain competitiveness. For manufacturers, moving to the cloud might also bring a number of difficulties. This blog article will go over some of the most prevalent cloud transformation-related manufacturing issues and supply chain solutions (intwo, 2024).

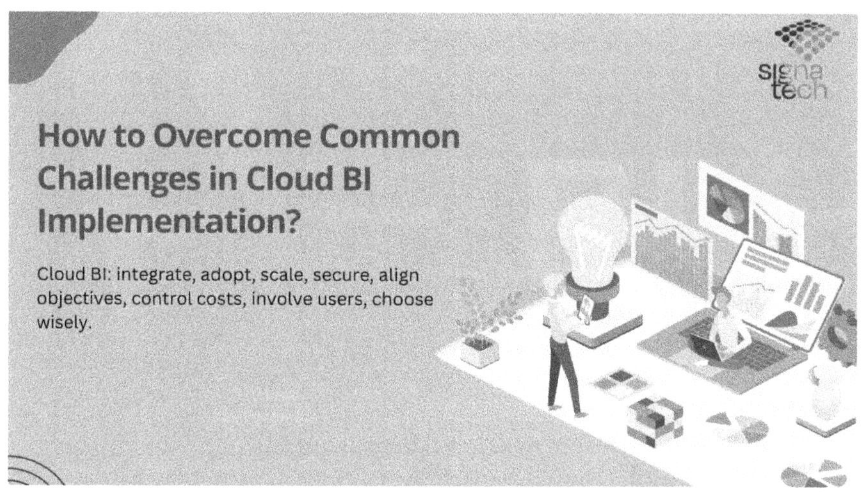

Figure 5.3: How to Overcome Common Challenges in Cloud BI Implementation?

Source: *- (Signatech, 2023)*

1. **Data security: Protecting confidential information in the cloud**

The primary concern for manufacturers when moving to the cloud is the security of their sensitive data. Many manufacturing firms work with sensitive information including product specifications, customer details and lots of financial information. It becomes very risky for any company if this information falls into the wrong hands or is lost in any way possible. The only way to reduce these dangers is to verify that cloud providers are using security measures like encryption and multi-factor authentication. Frequent inspections and vulnerability assessments, including security penetration testing, should also be conducted in order to identify these weak areas. Additionally, security is a relationship concern for manufacturers who may be required to adhere to regulations like GDPR, PCI-DSS, and HIPAA. For the manufacturers relevant to these regulations, they must assure themselves that their cloud provider meets those requirements and can produce the necessary paperwork and accreditations (intwo, 2024).

2. Seamless integration: Overcoming challenges of migrating legacy systems to the cloud

The ability of current systems to integrate with the new cloud systems is the other issue that manufacturers voice while making the move to the cloud. The majority have legacy systems, which are the foundation of their business and have been in place for months or even years. The process of replacing these systems Howbeit, can be cumbersome and expensive. To overcome this challenge, the manufacture should go for a cloud provider that can solve integration services so that it will be able to interconnect with their local systems. These solutions can be data migration services, APIs or Application Programming Interface and middleware solutions. Manufacturers should also consider the hybrid approach which enables the existing systems to remain firmly in the factory while allowing the use of cloud services.

3. Scalability and resource management: Optimizing cloud capabilities for fluctuating demand

This situation of change is typical for manufacturing companies, which often have unstable demand for their products; here, it is obvious that calculating the necessary resources in advance can turn into quite a problematic question. It is observable that cloud-based systems are flexible which means that a manufacturer can accommodate himself to the program or expand when the need be. But this also pose some problems if the manufacturers are not capable of managing their cloud resources efficiently. To counter this, the manufacturers should register with a cloud provider who provides the automatic scaling and monitoring services. They should also agree on how to manage resources in the cloud and come up with policy and procedures in case of high access rates.

4. Cost management: Minimizing expenses in manufacturing cloud transformation

Cost is always a factor for manufacturing organizations and cloud transformation costs money. Merging data, converting systems, and educating the employees could get very expensive. Furthermore, manufacturers seem to face continuous costs of employing cloud service, including storage and computation charges. The best approach that manufacturing firms should consider in adopting cloud service is to opt for cloud service providers that adopt a usage-based billing model since it can reduce on overhead costs. They should also set up an expenditure plan for cloud transformation and check expenses against this plan regularly.

5. Training and support: Empowering employees for effective cloud adoption

Last but not least, there is training of workers who are required to work within the framework of cloud-based systems. The level of familiarity with modern technologies and equipment or tools can only be low and the employees will have to be trained on what to do. An effective cloud provider for manufacturers must include training and support services for their subscribers. They should also put measures in place for developing an independent program to train its human resource and support them as they master these new systems (Gangwar, 2017)this article presents an integrative research model that links, environmental, organizational and technological capability constructs. And to fill a literature gap, the study focuses on post adoption stages, i.e., actual usage and value creation, by surveying403 manufacturing firms in India. The data are analyzed through exploratory and confirmatory factor analyses, yet structural equation modeling further tests the proposed model. The results show that business, human and technological capital, change management, organizational culture, and regulatory and supplier support are all vital antecedents of cloud computing usage, with firm size moderating actual usage and performance. The unique insights the results provide can help organizations enhance their cloud computing adoption and thereby improve their business performance.",”author”:[{“dropping-particle”:””,”family” :”Gangwar”,”given”:”Hemlata”,”non-dropping-particle”:””,”parse-names”:false,”suffix”:””}],”container-title”:”Human   Systems Management”,”id”:”ITEM-1”,”issued”:{“date-parts”:[[“2017”]] },”title”:”Cloud computing usage and its effect on organizational performance”,”type”:”article-journal”},”uris”:[“http:// www.mendeley.com/documents/?uuid=9f4620e5-9ab6-4349-8829-647cb3406c0a”,”http://www.mendeley.com/ documents/?uuid=63d9a8f6-a82f-4fa9-bc17-50e34732fb48”]}],”m endeley”:{“formattedCitation”:”(Gangwar, 2017.

Thus, they conclude that manufacturers would gain many benefits from these cloud transformations, including efficiency, cost advantage, and scalability. However, it comes with a number of problems which need a solution as outlined below. When these issues are well understood and factored into the success strategy of manufacturers, they will realize a smooth transition to the cloud. In two, as a services provider in the cloud, delivers multiple services and support for manufacturing firms to tackle these issues and move to the reality of cloud opportunity. Call us for other information on how we will assist you in migrating your business to the cloud (intwo, 2024).

5.7 Case Study: Smart Manufacturing Enabled by Cloud BI

Artificial intelligence (AI), the internet of things (IOT), big data analytics, and other smart technologies are used in the current notion of smart manufacturing. Therefore, with the aid of these technologies, SM helps producers make the best choices, advancing production processes, cutting costs, and enhancing product quality(Bryan Christiansen, 2023).

1. Smart Manufacturing Use Case 1: Predictive Maintenance

One common use of the ideas of smart manufacturing is predictive maintenance. This method of equipment maintenance uses statistical methods and technology to predict when equipment may break down.

Instead of addressing major issues when equipment malfunctions, predictive maintenance allows the maker to address possible problems before they become serious ones. By addressing them beforehand, the company will be able to minimize output loss, which might be expensive.

Additionally, by using machine learning algorithms on this data, equipment defects may be identified and probable breakdowns can be forecasted. Manufacturers can use this knowledge to plan maintenance during periods of low output to minimize or prevent any disruptions.

Some of the benefits of predictive maintenance in smart manufacturing are:

- **Downtime Reduction:** Predictive maintenance lowers downtime, improves equipment dependability, and boosts overall productivity by anticipating and preventing equipment issues before they happen.

- **Maintenance Cost Savings:** Manufacturers may prevent needless downtime and the related expenses of emergency repairs by scheduling maintenance tasks at convenient periods.

- **Product Quality:** The possibility of problems brought on by malfunctioning equipment is decreased since the equipment is performing at its best.

Case Study

One of the biggest suppliers of industrial vehicles, including forklifts, worldwide is TVH. They began utilising IoT-enabled forklifts to do predictive maintenance in 2017. This helped the business optimize their technician scheduling and parts management by enabling them to remotely diagnose and classify repair incidents.

TVH observed a 30% decrease in maintenance expenses and a "significant increase in the uptime of the machines" as a result of the predictive maintenance program.

Smart manufacturing relies heavily on predictive maintenance, which is revolutionizing how manufacturers maintain their equipment and resulting in notable increases in productivity, profitability, and efficiency(Bryan Christiansen, 2023).

2. Smart Manufacturing Use Case 2: Quality Control

It is a very crucial process which enables development of products with right quality that fully meets customers' needs and requirements as well as legal requirements. The aim is to decrease the probability of business related to products and product recall affecting a brand.

Therefore, the smart manufacturing technology is a tremendously important move in quality control since it can be conceived as an opportunity for controlling and monitoring continuously the whole process. Manufacturing systems are smart in the way that they facilitate convergence of data from different sensors and other connected devices through the manufacturing process. The information can also be useful in problem-solving, troubleshooting and identifying issues that may be on the production line so that effort can be taken to control the process and reduce defects and non-conformities.

Besides, smart manufacturing, for example, through big data and artificial intelligence, including machine learning algorithms, is able to analyze the data, look for signs and inability to look for patterns and abnormality to allow for patterns such as the predictive maintenance of heater heads and other quality control measures.

Some of the benefits of smart manufacturing for quality control are:

- **Improved Quality:** Therefore, manufacturers can prevent flaws which may jeopardize quality of the final item and customer satisfaction.

- **Risk Reduction:** Smart quality control also reduces the possibility of products that could lead to recalls which not only gives brands a big blow on expenditure but their image too.

- **Efficiency:** Data gathered in real time may be used to make optimized manufacturing processes as well as minimize wastage levels.

Thus, when the smart manufacturing technology has stably emerged it is believed quality control will turn out to be one of the factors that can define the success and the competitiveness of the new form of manufacturing industry.

Case Study

A large adhesive manufacturer sought to get a lot better at its quality management. The managers believed that there generally existed improvement opportunities.

Thus, they carried out an evaluation that showed a lack of understanding of how fragile the quality assurance parameters are and their relation to output quality. They also identified prospects for the next level real-time monitoring and process controls. Those weaknesses imply the possibility of variations in what contributes to high inventory, failure to meet customers' deadlines and leading directly to many severe financial consequences.

They provided a qualitive platform which was also capable of implementing and analyzing the historical information apart from regulating the segment. The model, they said, gave them a rather good impression of the exact parameters where more important when it came to quality. Thus they could dial-in on the most critical variables of their processes.

3. Smart Manufacturing Use Case 3: Supply Chain Optimization

In gaining familiarity over smart manufacturing, corporations consequently acquire data from assorted facilities in deliverer chains in real time. Such a system is beneficial to them in that it helps them to make improved decisions concerning production, stocking and delivery.

Cost reduction has singled out many aspects whereby organisations can reduce the cost of their products and therefore the cost of setting up new products must address the following points, Smart manufacturing also plays its part in the overstock or facing of the raw materials in managing stocks in the best way as possible. The idea is that the working capital of distinct companies can be well managed and resources can be well channeled.

The second advantage of FRS 15 is better flexibility. A special feature of solving problems in the supply and demand model lies in the fact that the condition in the market or in relations between supply and demand may be corrected without delay with the help of analytical data. Smart manufacturing all enables an organisation to effect change to the manufacturing process timely, and this results into a short time to respond to customer needs, short lead time, shorter delivery turns.

Case Study

A business aimed to increase its market share while decreasing operational expenses; it blamed its supply chain for the problem. In order to examine the selected supply chain network, the second and third quantitative components include the use of an IoT platform with analytics.

With the use of sensors put into a few discontinued items, they developed an asset tracking system. The staff were able to locate the merchandise more quickly and get it ready for delivery at the correct time thanks to that solution, which allowed the company to run more efficiently.

Second, the data collected by sensors was then transmitted to applications and panels to provide highly specifics real-time information on manufacturing and shipyard. In doing so, the business was well placed to monitor on the status of the work in progress inventory and factories.

The company restricted the unnecessary motions therefore increasing labour productivity by 3-4 percent and reducing total finished stock. In addition, signaling led to a two-day improvement in the pick-up time to the shipping function. Scheduling was also improved where concerns delivery to dealers, thus improving the overall business relation and customers.

A Final Word on Smart Manufacturing

When put into practice, smart manufacturing technologies like supply chain management, quality control, and predictive maintenance can greatly enhance output. Manufacturers may eliminate waste, achieve solid price economies, and boost effectiveness and profitability through the transparent integration of real-time data and advanced analytics (Bryan Christiansen, 2023).

All these use cases also make it possible to have anticipative measures, which are capable of avoiding probable problems, hence, good results in customer satisfaction and loyalty. The world of manufacturing is changing rapidly and thus smart manufacturing technologies are getting critical success factors that help the industry keep up with the world's innovation.

5.8 Chapter Summary

Data analytics is changing the arena of manufacturing because it enables the conversion of large amounts of raw data into meaningful information of value for improvement of efficiency, productivity and decision making.

It leads to better quality, reduced costs and minimal time is spent on repair and maintenance of the equipment. It is a crucial component in this transformation because it provides real-time monitoring of equipment and anticipates failure, which saves costs, increases the machinery's useful life, and avoids interruptions to production. Quality is also improved since the early detection of defects helps reduce wastage, and maintain product quality on the market. In supply-chain data analytics enhances demand planners, inventory and distribution so that there is timely delivery at a lower cost.

This paper identifies four key enablers of implementing predictive maintenance: goal alignment, performance metrics, data acquisition, and advanced technology. Training efforts are required in data interpretation and sensor systems, maintenance processes must integrate and plan efficiently, and there must be dedication to the concept of kaizen. A real-time data streaming, IoT integration also enhances manufacturing analytics by guaranteeing timely decision-making. APIs, middleware, cloud, edge computing maintain data/control flow, secure processing, and actionable analytics meet international standards.

The use of technology in cloud analytics is very important in quality assurance since it produces timely and real results which reduce on defect rates and wastage while conforming to the set standards. Nevertheless, issues such as data security concerns, issues related to the migration of current systems, and achieving cost efficiencies require strong modes of security, integration services, and conditions of cloud computing. Smart manufacturing which is inspired by intelligent technology applications such as AI, IoT and big data is mainly driven by smart solutions ranging from predictive maintenance, real time quality controlling to efficient supply chain management. These developments improve the operational quality of manufacturers in aspects of cost cut, productivity and satisfaction which tactfully positions manufacturers in the frontier to meet up with the increasing data driven market.

Multiple Choice Questions (MCQs)

1. **What is the primary goal of implementing Cloud BI in manufacturing?**

   a. To reduce employee workload

   b. To improve production efficiency and decision-making

   c. To reduce costs without improving quality

   d. To eliminate the need for physical factories

2. **Which of the following is an example of a strategy used to improve production efficiency through data analysis?**

   a. Hiring more staff

   b. Monitoring real-time data from production lines

   c. Decreasing production speed

   d. Limiting product variations

3. **Predictive maintenance strategies in manufacturing aim to:**

   a. Replace machinery more frequently

   b. Predict equipment failures before they occur

   c. Increase the number of machines in operation

   d. Focus on reducing employee downtime

4. **How does IoT data integrate with Cloud BI solutions in manufacturing?**

   a. By collecting data from machines and sensors in real-time

   b. By reducing the need for any data collection

   c. By replacing the use of cloud-based systems entirely

   d. By limiting machine performance data collection

5. **Which of the following is a key benefit of Cloud BI for quality control in manufacturing?**

    a. Reducing product variety

    b. Automating the production process entirely

    c. Enhancing the ability to detect quality issues early

    d. Eliminating the need for human workers

6. **What is one of the major challenges in implementing Cloud BI in manufacturing?**

    a. High energy consumption

    b. Lack of available data

    c. Integration with existing legacy systems

    d. Limited access to cloud services

7. **What is the purpose of a case study on smart manufacturing enabled by Cloud BI?**

    a. To highlight the challenges of traditional manufacturing methods

    b. To showcase how Cloud BI can drive improvements in manufacturing processes

    c. To show the benefits of reducing data analysis

    d. To explain how manufacturing can work without cloud technologies

8. **Which of the following is a key advantage of using Cloud BI for manufacturing companies?**

    a. Ability to scale data processing capabilities as needed

    b. Reduced need for data security measures

    c. Elimination of all manual processes

    d. Reduced reliance on technology

9. **In the context of Cloud BI, what is meant by 'quality control enhancements'?**

   a. Increasing production speed at the expense of quality

   b. Using data analytics to monitor and improve product quality

   c. Focusing on reducing the number of defects in production

   d. Using IoT to monitor machine temperature

10. **Which of the following would most likely be a part of overcoming challenges in manufacturing Cloud BI implementation?**

   a. Expanding physical infrastructure

   b. Ensuring seamless integration with existing IT systems

   c. Reducing employee engagement

   d. Decreasing cloud storage capacity

Answer

1	2	3	4	5	6	7	8	9	10
b	b	b	a	c	c	b	a	b	b

Chapter 06

FINANCIAL SERVICES AND CLOUD BI INTEGRATION

6.1 Chapter Overview

The use of Cloud Business Intelligence (Cloud BI) in financial services industries is changing the way business entities approach risk management, fraud detection as well as compliance with regulation. This chapter describes how cloud-based analytics provides innovation in segmentation, profitability, and reporting alongside the customer challenges of security and privacy. It is evident from Cloud BI that financial organizations can obtain up-to-date information and promote changes in decision-making and efficiency. Using a detailed case of one of the biggest international banks as an example, the chapter demonstrates how the Cloud BI solutions were initially implemented and how it is evidence of the dynamics that cloud technologies can bring to the financial industry.

6.2 Risk Management Using Cloud-Based Analytics

The process of identifying, assessing, and mitigating the risks associated with cloud computing services is known as cloud risk management.

Companies use proactive risk management techniques to shield their infrastructure, data, and apps from possible dangers (Matt Pacheco, 2024).

The Importance of Cloud Risk Management

Cloud risk management may help businesses save money, improve their compliance procedures, lower the risk of data breaches, increase business continuity, and maintain their brand. Whether or not you choose to use cloud risk management might determine whether your organisation survives or fails.

1. Potential Consequences of Inadequate Risk Management

Neglecting to control cloud security threats can have a number of negative effects, including:

- **Data breaches:** Data breaches have the potential to seriously harm your company. On average, they cost organisations $4.45 million. Furthermore, cloud-stored data was a part of 82% of breaches.

- **Financial losses:** Financial losses linked to a data breach include lost revenue, labour expenses for detecting and fixing the breach, fines from the authorities, legal fees, and more.

- **Compliance issues:** Certain data privacy laws must be followed by specific industry. You might be subject to regulatory repercussions if your company has insufficient safeguards in place.

- **Reputational damage:** Although cloud risk management may enhance your image, neglecting to consider possible risks can lead to a tarnished reputation and a decline in trust. There are certain clients who might never return.

2. Common Cloud Security Risks

The first step in cloud risk management is to identify typical security threats to the cloud. You can create a risk management plan that will best safeguard your data and infrastructure by knowing what may go wrong for your company.

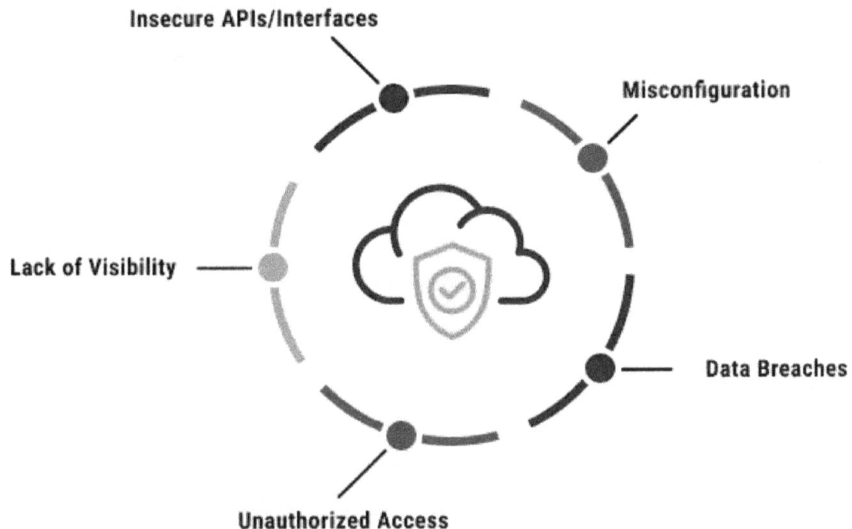

Figure 6.1: Common Cloud Security Risks

Source: - *(Matt Pacheco, 2024)*

- **Misconfiguration:** A major contributor to data breaches, cloud security settings are the cornerstone of your cloud infrastructure. Even minor setup errors might result in serious risks, such exposing private information or making it simpler for illegal access to enter.

- **Data Breaches:** The amount of data that may be kept in cloud systems makes them appealing to hackers. You can be exposing your data to intrusion if you are unaware of your security obligations with regard to your cloud provider (more on the shared responsibility model below).

- **Unauthorized Access:** In 2023, phishing and credentials that had been stolen or compromised were the two most common attack methods. After obtaining credentials, hackers can swiftly infiltrate other systems and use ransomware to prevent the rest of your company from accessing any or all of your data and apps.

- **Insecure APIs/Interfaces:** The functioning of cloud apps is dependent on application programming interfaces, or APIs. However, inadequately protected APIs can lead to vulnerabilities that malicious actors can use to access private information or affect how well your systems work.

- **Lack of Visibility:** If you are unable to observe your full cloud environment, it may be difficult to comprehend your security threats. Businesses might not be able to efficiently monitor their cloud environments and spot suspicious activity if the proper procedures and tools aren't in place.

How to Assess and Mitigate Cloud Risks

Assessment and reduction of risks follow identification. By employing the appropriate tools and developing the appropriate models, your company may lower cloud security risks.

Strategies to Assess and Mitigate Cloud Risks

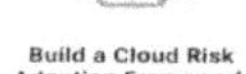

Figure 6.2: Strategies to assess and mitigate cloud risk

Source: - *(Matt Pacheco, 2024)*

1. Study the Shared Responsibility Model

Under a shared responsibility model, users are mainly responsible for ensuring the security of their cloud data, with some help from the cloud provider. Responsibility levels are defined by the choice of SaaS, PaaS, or IaaS delivery model. For example, while AWS may be responsible for safeguarding its global infrastructure and the applications that run on it (data, networking, storage, and computation), users are ultimately accountable for protecting various aspects of their own systems, including operating systems, firewalls, and personal information.

2. Build a Cloud Risk Management Framework

A solid foundation for managing cloud risks can be laid after you have a firm grip on your duties in the cloud. The steps to follow in order to make sure everything goes according to plan are as follows: find potential threats, figure out how bad they could be, devise ways to lessen their impact, write reports, and implement risk governance.

3. Perform a Cloud Security Assessment to Identify Risks

Verify the proper configuration of data security, compliance processes, and access controls by conducting a cloud security assessment based on the development of your framework. You can utilise tools and do penetration testing, which mimics an attack, to ensure your risk management techniques will work as expected.

4. Leverage Tools and Services to Mitigate Risks

Your business's anticipated security risks will also play a role in determining the best security solutions and services to implement. Security technologies that you would like to incorporate include features like encryption, security monitoring, and access control. There may be some that are proprietary to cloud providers and others that are third-party solutions.

- **Firewalls and Intrusion Detection/Prevention Tools**
  - **AWS:** AWS WAF and Shield for web applications
  - **Azure:** Azure Firewall and Azure Sentinel
- **Logging and Monitoring for Security-Related Events**
  - **AWS:** CloudWatch
  - **Azure:** Azure Monitor and Azure Security Center
- **Integration with SIEM Tools**
  - **AWS:** CloudWatch can be integrated with third-party SIEMS, like Splunk
  - **Azure:** Azure Sentinel supports third-party integrations and has built-in SIEM capabilities
- **Incident Response and Real-Time Threat Detection**
  - **AWS:** Guard Duty
  - **Azure:** Azure Advanced Threat Protection

5. Continuously Monitor and Assess Risks

Risk management is a continuous process. The panorama of threats is always expanding and shifting. You will need to keep an eye out for fresh threats and vulnerabilities in order to safeguard your cloud environment. To ensure that your cloud environment is not only safe now but will also be safer from attacks in the future, schedule regular security evaluations and allocate resources for continuous monitoring.

6.3 Fraud Detection and Prevention with Cloud BI

The emergence of digital payments and the dissolution of branch networks have created new difficulties for financial institutions as the global banking environment changes. Fraudsters are always coming up with new and creative ways to steal from banks and their clients.

In 2022, bank transfer or payment fraud cost the United States an astounding \$1.59 billion, according to data from Statista (korcomptenz, 2024).

The banking industry is now threatened by a number of things, most notably fraudulent operations such as identity theft, account takeovers, and payment-related fraud. The need for improved cloud-managed fraud protection systems is underscored by the over 1.7 million identity theft allegations that the Federal Trade Commission received in 2021.

Payment Fraud is a Serious Challenge

Payment fraud costs businesses a lot of money in addition to harming individual consumers. According to projections, losses from credit card theft are predicted to exceed \$49 billion by 2030, indicating a concerning pattern of ongoing rise.

Financial institutions are using cloud-based fraud detection and prevention systems in order to successfully mitigate these risks.

Enhancing Security with Cloud-Based Solutions

The fact that over half of respondents to a Global Banking Fraud Survey recovered less than 25% of their losses from fraud occurrences shows that financial institutions need to make fraud prevention a top priority. Because of this, implementing cutting-edge cloud-based solutions becomes essential.

Leveraging the Cloud, Advanced Fraud Detection and Prevention Systems Provide:

1. **Real-time Monitoring and Fraud Alerts**

   Real-time tracking of consumer behaviour and purchases is made possible by cloud consulting solutions for banking fraud detection. In a couple of seconds, the analytics apps use a specific algorithm that examines vast amounts of data to find anomalies and suspect activities. When the system detects a possible fraud event, it generates alarms to help banks stop more fraud incidents and losses.

2. Machine Learning and Artificial Intelligence

Automated fraud detection in the cloud learns from past events through with machine learning and artificial intelligence algorithms. Such systems analyze past experiences, use the information found there, and modify fraud detection in banking models to new threats. With the help of AI solutions, the frequency of false positives and false negatives in the case of fraud is eliminated, and the overall volume of the customers' transactions is not affected.

3. Enhanced Data Security

As for the data breaches, where the data was stolen or leaked intentionally or unintentionally, it was recorded that in 2021 the cases amounted to 1,862 and this is 68% more than in 2020. Solutions that operate on cloud infrastructure are fortified with measures for protecting financial data. Since cloud service providers have access to large amounts of customer information, they never take this issue lightly; users' data is encoded and protected from access by unauthorized individuals and undergoes security checks at all times. Also, cloud systems have the emergency features of disaster recovery and back-up of data in case of system or calamity failure (korcomptenz, 2024).

4. Scalability and Flexibility

Fraud detection services that are hosted in the cloud help banks offer services as per the need of the market. Due to scale variability, the systems used in cloud infrastructures can immediately alter the volume of connection resources to accommodate the ever-increasing number of customer transactions and information. These scale factors make certain that banks can continue to preserver fraud detection measures without a significant impact on system performance.

5. Collaboration and Compliance

cloud-managed networks enable the sharing of knowledge and resources among the financial industries. When multiple data sources are combined, the banks can be in a position to identify fraud trends and patterns across the banking institutions. Such collaborative and combined approach enhances fraud defenses and allows for the prevention of new threats as they emerge. It may also explain that purposes, employing anonymized data and insights might positively influence the banking sphere as a whole to present a united front against fraud.

Moreover, cloud-based solutions incorporate many compliance features which go a long way to helping banks in their compliance with the various requirements. Many of these solutions include functionality to track compliance with the rules and regulation irrespective of the model including the Payment Card Industry Data Security Standard (PCI DSS) as well as the General Data Protection Regulation (GDPR).

6. Cloud Computing to Mitigate Risks Effectively

According to McKinsey & Company cloud computing reforms are opening opportunities and prospects of thorough risk management especially in the banking industry. This paper established that security features offered by cloud computing provide solutions to different types of risks covering financial risks such as market risks, credit risks, liquidity risks and non-financial risks such as fraud risks and cybersecurity risks, financial crime risks among others.

In turn, cloud services might help minimize the time for risk teams to respond to potential security breaches while avoiding large capital expenses simultaneously.

7. Cloud-Based Tools to Stop Economic Crimes

McKinsey gives an example of a bank that uses cloud consulting services in a unique way to stop money laundering and other illegal activities. Thus, by assisting the risk management teams in identifying suspicious transactions that were previously assessed solely by human eyes, the bank is able to establish a connection between the individual and the company using Global Social Network Analytics. Additionally, cloud services may help detect data breaches and track down the criminals responsible for them.

6.4 Regulatory Compliance Reporting in the Cloud

When it comes to the security and privacy of cloud computing, cloud compliance is making sure that all data and services stored there follow the rules. Achieving compliance with these legal requirements is the objective of cloud data management, storage, and security. The necessary regulations may include healthcare-specific statutes like HIPAA or more general data protection standards like GDPR (aquasec, 2023).

There is more to cloud compliance than just reviewing the items on a compliance checklist. The stringent regulatory compliance inspections, service audits, and cloud service updates to comply with the new legislation all occur concurrently with this cycle. Thus, cloud compliance seeks to guarantee that the cloud supports privacy, data security, and business continuity.

Practical compliance measured within a cloud environment can assist your business to operate without fines and penalties for non-compliance. But above all else, it is about how to share information with your customers, defend your business image and guarantee sustainable enterprise performance:

- **Taking security seriously:** The assurance that a company takes data security and privacy seriously is provided by cloud compliance. In today's corporate environment, this can help you stand out, especially in fields where data security is crucial.

- **Improving risk management:** Cloud compliance plays a significant part in risk management as well. You may prevent data breaches, detect and reduce any security threats, and guarantee business continuity by following compliance requirements.

8 Key Compliance Standards and Regulations in the Cloud

There are a number of important compliance guidelines and rules that cloud-based enterprises must understand. A summary of some of the more significant ones is provided below. (aquasec, 2023)

- **PCI-DSS**

    Businesses must ensure a secure environment when receiving, handling, storing, or transmitting credit card data in order to comply with the Payment Card Industry Data Security Standard (PCI-DSS). If your business handles customer credit card information, PCI-DSS compliance is mandatory regardless of the size or frequency of your transaction volume.

    Cloud storage and processing of credit card information must adhere to the PCI Data Security Standard (PCI-DSS). Businesses that use cloud computing or provide cloud services must ensure they have secure networks, implement strong access control policies, test and monitor their networks often, and stay up-to-date with vulnerability management programs to be compliant.

Figure 6.3: PCI-DSS compliant logo

Source: - *(Isms, 2022)*

- ### ISO 27001

  ISO 27000 is the family and ISO 27001 is the standard on information security management system. This offers a guideline on how to set, operate, maintain and systematically enhance an information security management system. By implementing the ISO 27001, a company proves that it addresses management of company's sensitive information in a more systematic manner than it is in a case where the company has not embarked on the compliance process.

  Regarding the cloud, ISO 27001 enables organizations to set an all-encompassing ISMS. Organizations that opt for cloud service providers who are ISO 27001 certified can as well demonstrate that their suppliers possess safe systems that would ensure the privacy, confidentiality and accessibility of data. Conversely, the organizations that hired the cloud services must make sure that the providers they have chosen adhere to these standards to aid in the general protection of data.

Figure 6.4: ISO 27001 LOGO

Source: - *(Zeotap, 2024)*

- **SOX**

  A law in the United States called the Sarbanes Oxley Act requires all publicly traded corporations to adhere to specific business standards in order to guarantee the accuracy of their financial reporting. It contains features for using a cloud environment to store, process, and retrieve financial data. This essay makes the case that businesses that deal with financial data and stock market corporations should adhere to SOX.

  The regulation which is associated with cloud computing is the Sarbanes-Oxley Act also known as SOX, which is related only to the situation when financial data is being stored and processed in the cloud. In general, the law demands that the companies must have ensured control like cryptography and access control to safeguard the relevance and integrity of the information put in cloud databases. Moreover, cloud service providers need to provide certain features and utilities to customers for their SOX compliance needs like facility for audit and data backups.

Figure 6.5: SOX logo

Source: - *(Hyperproof, 2022)*

- **NIST**

  The National Institute of Standards and Technology (NIST) is a government agency that issues guidelines and standards for the nation's IT infrastructure. However, a large number of private-sector businesses worldwide have also adopted these standards. The NIST framework offers a logical way to handle cybersecurity risk management.

  Because it provides a set of protections that organisations and CSPs may apply to secure computer systems against risks like cyber-attacks that could harm cloud data, the NIST framework is utilised in cloud computing. Regarding the methods for risk identification, assessment, prevention, and supervision in cybersecurity problems, these suggestions are beneficial for the third component, which is developing a sufficient risk management strategy in a cloud environment.

Figure 6.6: The National Institute of Standards and Technology (NIST)

Source: - *(johns hopkings, 2024)*

- **GDPR**

   The General Data Protection Regulation (GDPR) is one of the EU's privacy and data protection rules. It is one of the strongest laws in the world for protecting the private data of EU citizens, and it applies to all businesses, no matter where they are based, that handle the personal data of EU citizens.

   To that extent, the GDPR lays down that any CSP, or any business, that uses CSPs for storing or processing personal data of EU citizens have to have adequate data protection controls in place. Some of these requirements include among others the requirement to ensure that data is encrypted, that data should be obligatory deletable and that the exhibition of data breaches is compulsory.

Figure 6.7: GDPR logo

Source: - *(Loginradius, 2024)*

- **CCPA**

  The CCPA is a state statute which aims for improvement of privacy protection and consumer rights in California. CCPA applies to any for-profit entity that derive revenue from selling consumers' personal information or business that purchase, sell or share consumers' personal information, except for the personal information that is publicly available or obtained from the consumers themselves.

  In regard to cloud computing, organizations that obtain consumers' personal information of California residents and maintain or process it in the cloud must meet the CCPA privacy standard. This is by ensuring that the security protocols are set correctly, consumers are informed about their privacy, and the organization should have the ability to fulfill consumer request of access, deletion, and no sells of their data.

Figure 6.8: CCPA act

Source: - *(Sansico, 2019)*

- **HIPAA**

  The Health Insurance Portability and Accountability Act (HIPAA) adopted acts the benchmark of sensitive patient data protection known today as the HIPAA. In handling protected health information (PHI), business entities have to check and double check if all the physical, network, and processes have implemented and enforced all the necessary security features.

  Thus, it is inferred in the context of the Cloud as it relates to HIPAA that cloud service providers and businesses that use cloud systems for processing or storing PHI should have robust network, procedural, and physical security safeguards. Additionally, BAAs that must be agreed between cloud providers and healthcare providers to ensure secure PHI management are underlined.

Figure 6.9: HIPPA compliant

Source: - *(Ware, 2024)*

- **FedRAMP**

  A government-wide initiative, the Federal Risk and Authorization Management Program (FedRAMP) focusses on the security evaluation, authorization, and monitoring of cloud services and products that are often utilised by federal agencies.

  In connection with cloud computing, FedRAMP defines the security assessment process, authorizes cloud products and services used by federal agencies, and monitors them continuously. Businesses utilizing FedRAMP services need to meet extremely high standards of security that have been specifically assessed by the FedRAMP authority, and thus if they are approved for use, federal agencies can feel more at ease when adopting their cloud services.

Figure 6.10: FedRAMP logo

Source: - *(FedRAMP, 2020)*

7 Steps to Achieving Cloud Compliance

This is important because to achieve cloud compliance an organization has to go through a systematic approach that may involve several significant steps:

1. **Implement a Shared Responsibility Model**

   The very first approach in adopting cloud compliance is through the shared responsibility model. In reality, the shared responsibilities as to the security and compliance obligations are divided between the cloud service provider and the customer. However, the degree of responsibility depends on which cloud model is in place: IaaS, PaaS, SaaS.

In theory, the security and compliance of the cloud environment—that is, the actual servers, networks, and databases—must be assumed by the cloud service provider. On the other hand, the customer is responsible for user management, data security and compliance, and cloud-based data. By helping the parties understand their individual obligations to prevent non-compliance, the shared responsibility model guarantees that compliance responsibilities are visible.

2. Implement a Governance Framework

In order to revitalize governance frameworks, institutions, and cultures, the next stage is to promote good governance practices. The collection of instruments that control and monitor the condition of the cloud environment is referred to as governance in the context of cloud compliance. In order to provide guidelines for cloud operations that are in keeping with best practices and minimal regulatory criteria, a governance framework may occasionally be referred to as a cloud operating model.

The major elements of strong cloud governance framework are risk management, policy management and change management elements. Cloud risk management comprises recognition, evaluation and mitigation of risks that are relevant to the running of clouds. Policy management involves the formulation of policies for governing cloud operations and the putting in place and the application of policies for governing cloud operations. It is the process by which management oversees modification to the cloud environment to avoid interference which is compliance.

3. Develop a Compliance Strategy

Finding a broad compliance plan is the last stage that follows the implementation of a shared responsibility model and a sound governance framework. Every action and activity required for cloud compliance assurance and compliance is outlined in a cloud compliance framework.

A clear and specific statement of the organization's compliance goals and objectives, a description of the compliance standards and regulations that apply to its operations, the assignment of compliance responsibilities, and a thorough explanation of the compliance procedures and processes are all part of the process of creating a compliance strategy. Additionally, it should include a backup strategy for resolving non-compliance issues and evaluating compliance performance.

4. Deploy Compliance Tools and Controls

Cloud compliance involves the utilization of all the necessary tools and compliance controls that enable cloud compliance outcomes. The discussed technologies enable compliance tasks to be performed automatically with minimal human intervention, to monitor compliance events and violations, and to produce reports on compliance status.

Data protection tools, compliance management software, and security information and event management systems are some of the fundamental components of compliance tools. Tools for recording, organizing, and reporting compliance activities as well as those that notify users of current or possible noncompliance concerns are examples of compliance work management tools.

The procedures used in an organization's cloud environment to guarantee adherence to corporate and regulatory regulations and standards are known as compliance controls. Technical controls include firewalls and encryption, while administrative controls include audit trails and user access control.

5. Regular Auditing and Reporting

To ensure good cloud compliance, another important procedure that must be followed is auditing and reporting. Conducting audits helps to guarantee that the cloud environment only conforms to the criteria established by legal regulations. Additionally, they assist in identifying compliance concerns and areas for improvement.

The practice of providing the auditors' conclusions following an audit for the benefit of certain parties is known as reporting. These reports may be utilised for decision-making and future planning since they provide information on the compliance position and level of compliance.

6.   **Maintain Documentation**

Additionally, documentation serves to record compliance procedures and activities. They include accounting records, permit records, evidential audits, rules, procedures, and guidelines.

Documentation facilitates internal knowledge exchange and the recording of continuities in addition to helping the organisation demonstrate compliance to the regulator and external auditor. By doing this, a circumstance where some staff members, managers, or even auditors may find themselves in a situation where they were unaware of any compliance need or process is avoided.

7.   **Continually Monitor and Update Compliance Measures**

Lastly, achieving cloud compliance is a continuous process. It entails ongoing compliance measure monitoring and improvement. Despite the constantly changing regulatory environment and technological advancements, this ongoing procedure ensures that the cloud environment remains compliant.

Monitoring entails routinely assessing compliance performance and status, whereas updating entails modifying and enhancing compliance measures in response to regulatory, standard, and business needs changes as well as monitoring results. (aquasec, 2023)

6.5 Customer Segmentation and Profitability Analysis

Customer segmentation analysis?

Customer segmentation analysis is what is done when an organization is looking for the characteristics that will differentiate specific customers. This process if used by marketers and brands to understand which campaigns, offers or products to employ each time they are conducting a communication campaign with a particular segment.

For example, a retail brand seeking to define how to respond to customers who have not shopped in his/her store for the last one month, might define a segment of customers who made a purchase at his/her store but has not visited the e-commerce store at all in the last 1 month. Consumers in that segment could also be further segmented to see what kind of products these customers have been buying, their propensity to use a discount and so on. With all this information in hand, the marketing team is in a position to know which campaign to develop to reawaken these dormant customers. (Optimove, 2024)

Likewise, a firm can calculate and analyze in relation to customer segmentation the value of distinctly segments by looking at a segment's predicted Future Value, average order value, loyalty tier distribution, and more.

Customer segmentation important?

Customer segmentation has the advantage that marketing can tailor its approach to address every one of them. Given the volume of information available concerning customers (and many potential customers), a customer segmentation analysis enables the marketer to pinpoint separate customer segments with a high measure of reliability in terms of appropriateness of any one measure, demographic, behavioral or otherwise.

As the strategic objectives of a marketer often include getting as much value (sales and/or profits) out of each customer, it is useful to find out ahead of time how any specific marketing activity will affect the customer. Ideally, such "action-centric" customer segmentation will exclude any notion of the marketing action's near-term value, but consider instead the total customer lifetime value (CLV) that such an action will bring. Therefore, there is a need to cluster, or categorize, customers based on CLV.

- **CLV-Focused Customer Segmentation**

Of course, it is always easier to come up with assumptions and rely on 'intuition' to specify rules that would help and segment customers into meaningful categories such as those coming from a certain source, residing in a certain geographical area, or having purchased a certain product or service. However, these high-level categorizations will rarely afford these desired results.

Some customers are going to spend more compared to others and when that company relationships those customers, those are the customers that will spend most. Returning and quality customers will do so, for many years they so spend a lot. A good customer will spend less money frequently, or will spend more money for a short time. Others won't spend too much and others won't spend too long.

The correct way to approach segmentation analysis is the segmentation of customers according to the future value understanding to the given firm where the intention of the segmentation is to communicate to each group (or unique customer) in a way that will realize the highest potential value or lifetime value.

Approaches to Segment Customers

To define your customer segments once you have decided on the choice of categories and attributes, you then need to decide on the method you will employ to create those segments. According to classifications, there are basically two widely used Company Customer Segmentations namely the rule-based Customer segmentation and the Cluster Customer Segmentation.

- **Rule-Based Segmentation:** Establishing thresholds to decide which category a customer belongs to is the aim of rule-based segmentation. The method uses a set of principles to categorise clients into groups. Although the rule-based segmentation strategy is a straightforward method of classifying customers, it necessitates that you choose the characteristics that are utilised to divide up the client base each time. By showing the amount of clients in each group, it facilitates trend tracking. The rule-based segmentation strategy necessitates a lot of work to keep segments updated if customer behaviours change because it is hard to add new qualities.

- **Cluster-Based Segmentation:** Cluster-based segmentation finds the most effective approach to separate customers into groups that are as equal as feasible, as opposed to utilizing thresholds. In order to classify the data points into customer segments, it illustrates the relationships between them. These groups are produced via cluster-based segmentation using the K-means method. You can construct categories you were unaware existed and gain new insights from your data by using the cluster-based segmentation approach. Additionally, it can divide up its clientele according to a variety of criteria. Cluster-based segmentation is a challenging method to set up without a skilled data scientist, but it offers greater segmentation possibilities with minimal upkeep.

1. **Customer Segmentation Strategies**

   Effective customer segmentation strategies leverage data analytics and machine learning to precisely group customers based on behaviors, preferences, lifetime value, etc. Here are some examples of customer segmentation strategies:

2. **Behavioral Segmentation**

Understanding how customers behave is one of the most powerful ways to segment them. By analyzing purchasing behavior, website interactions, product and category preferences, and responses to past marketing campaigns, brands can tailor their messages and offerings to meet customer needs.

For example, brands can segment customers based on the day of the week and the time of day they usually shop to send promotions at these times. Alternatively, they can segment customers based on when they first registered, their last purchase/deposit date/amount, how frequently they engage, and so much more.

3. **Demographic Segmentation Combined with Behavioral Insights**

Traditional demographic segmentation, based on attributes like age, gender, or income, is still relevant but can be enhanced by combining it with other data points. For instance, using demographic data alongside psychographics or behavioral insights can provide a more precise view of customers. Optimove allows brands to integrate demographic insights with real-time behavioral data, providing a holistic view of the customer and enabling highly relevant, data-driven campaigns.

For example, a 35-year-old customer with a high income who frequently browses luxury products but rarely purchases, could be targeted with personalized offers that encourage and incentivize them to do so.

4. **Predictive Segmentation**

The rationale of the predictive customer segmentation is the use of state-of-the-art machine learning algorithms to detect patterns which may be concealed. In this case, back streams from previous years of the service can be used together with customer interactions' analysis to predict which segments are likely to convert, churn or will need special call to actions. This customer segmentation strategy allows brands to address customer needs to increase engagement, improve lifetime value, and drive revenue.

For example, predictive analytics models can identify trends and help marketers predict future behaviors, allowing them to deliver highly relevant and timely content. This could include targeting customers at risk of churn with personalized retention offers or rewarding loyal customers with exclusive deals to make them feel seen and appreciated.

Types of Customer Segmentation Models

There are numerous varieties of consumer segmentation strategies, even though it is advised to divide up your clientele according to their CLV. RFM segmentation, longevity, and segmentation by cluster analysis are a few of the more widely used kinds. To accomplish their goals, some marketers may even combine one or more segmentation strategies. Regularly updating new data and keeping an eye on dynamic changes are necessary for accurate client segmentation.

In order to begin segmenting the customer base, marketers must first create groups of customers, regardless of the type of segmentation model they choose to employ. As a result, marketers typically have a number of tiers for every kind of segmentation model. After that, marketers can combine various model tiers to produce more precise segments. For instance, marketers will have a segment of highly active, recently acquired customers if they combine the highest tier of customers based on an RFM model with a low longevity tier. (Optimove, 2024)

Businesses that use a customer segmentation model find it to be a useful tool for figuring out the best messaging and approach to employ when marketing to their target audience. It is perfect for conveying highly relevant and personalized messages, but it can also be used more strategically by revealing effective marketing techniques to boost customer lifetime value (CLTV) and retention.

Customer Profitability Analysis?

Customer profitability analysis enables you to divide up your client base according to how much money they bring in for your company and to focus your marketing, customer support, and operational expenses on the customer groups that generate the greatest revenue for your company.

Customer profitability goes well beyond customer lifetime value or even the gross or net margin made on a transaction. The profit (customer spend minus customer cost) from each interaction a customer has with your business, such as customer service interactions, returns, custom fulfilment expenses, and more, is known as customer profitability.(DAN LEBLANC, 2023)

Why Measuring Customer Profitability Is Important

The aim of any business is to be profitable. How achievable and sustainable it is depending on a number of elements, but your customers' profitability is the most important one. In addition to being vital, measuring client profitability may be illuminating.

The framework for conducting a client profitability analysis should be simpler to update once a year or as frequently as your company sees appropriate. Running one has the major benefit of allowing you to ascertain whether particular clients are truly costing you money instead of generating it.

In certain cases, you can discover that the client segment you haven't yet identified is more valuable to your business than the one you believed to be the most significant (most profitable), which indicates that your whole strategy has to change.

It typically boils down to service expenses if you're wondering how certain clients can be costing you money.

Do you have any consumers that contact customer service a lot? Do certain clients have particular fulfilment needs that need greater labour and fulfilment expenses? Or do you have consumers that use your free shipping and return policy excessively? Following a customer profitability study, you will begin to identify these segments.

The Steps in Customer Profitability Analysis

1. **Step 1: Define your customer costs**

   You need to know where your customers can interact with your organization and how much money you spend on operations before you can calculate client profitability. Customers may also be charged for additional costs beyond the price of the products or services they actually purchase, such as:

   - Marketing costs (e.g., your Cost Per Acquisition)

   - Customer service contact costs

   - Social media contact costs

   - Shipping costs (especially if you fund return shipping)

2. Step 2: Define your customer groups

   The topic of customer profitability segmentation will be discussed next. Depending on the size of the company or business unit they purchase from, some companies have clearly defined client categories. You can establish these client categories right now if you don't already have them defined.

   For instance, given your line of work, what kinds of clients do you have and why do they purchase from you? You may establish these groups even if you lack empirical support or a comprehensive third-party research since you are familiar with your company and its product or products: One excellent method for consumer segmentation is creating customer personas. These may be created using market research, poll data, user data (demographic data), and more.

   To divide up your clientele into distinct groups, you may also perform an RFM analysis. Recency, frequency, and monetary analysis, or RFM analysis, is a very thorough method of customer segmentation for customer profitability analysis. Using an RFM analysis, you may build customer cohorts based on three factors: the amount of money spent, the frequency of purchases, and the recentness of the customer's brand purchases.

A strong matrix for comprehending important client categories is the end result. This lets you decide which groups are more important to interact with strategically.

3. Step 3: Find the data

It's time to look for the specifics in Step 3, which may need you to put on your detective hat. To do this, you'll need assistance from colleagues in the Step 1 business sectors.

To begin with, do you have data? Data is likely in your possession, but is it easily accessible and tracked?

Finding every piece of information connected to every single consumer is unlikely. The Selling, General & Administrative Expense (SG&A) line item typically lumps expenses like customer service charges together as a single expense. Now, however, you may begin to support it with additional scientific data.

Here are some ideas to track down more detailed data:

Do you keep track of the client's interactions with customer service? You might be able to identify the client categories that could over-index in contacting you by looking at a sample of customer support enquiries. Why is the next question? These expenses' justifications—whether they be refunds, customer service interactions, or something else entirely—will be useful in the future.

When you break down marketing spend and cost per transaction by marketing channel and link them to certain client groups, you may uncover hidden gems in this data.

In the end, you will get at least a list of typical expenses for each activity, including:

- The cost of marketing to get an order (cost per order)

- Average number of customer support interactions for each order

- Average price per customer support interaction

- Average rate of return

- Average cost of shipping

4. Step 4: Putting it all together: a customer profitability analysis example

   Suppose you have two parts of your clientele: A and B. When you look at the income generated by each segment for your company (per transaction) at a high level, Segment B appears considerably more appealing:

	Customer Segment A	Customer Segment B
Avg. Revenue per Transaction	$75	$100

Source: - *(DAN LEBLANC, 2023)*

Using this information alone can make you want to change your approach, spending, and even the way you develop new products in order to better serve Segment B. Ultimately, they are worth 25% more than Segment A. That would be a wise strategy based on these figures.

What if, however, the customers in Segment B aren't truly superior? Here's where a customer profitability study may really help your company.

Suppose that Segment B clients have unique fulfilling needs that cause labour and fulfilment expenses to double. Despite all of that, they actually cancel or return items far more frequently than Segment

A clients. They don't seem as nice now as they did when this example began, do they?

	Customer Segment A	Customer Segment B
Labor, Fulfillment and Restock	$12	$25
Shipping Costs	$10	$30
Customer Service Costs	$6	$12
Total Costs	**$28**	$67

Source: - *(DAN LEBLANC, 2023)*

Even before the price of the product or service is taken into account, those Segment B clients are really costing you a lot of money. Overall, for the long-term viability and expansion of your company, you would be better suited directing your attention, strategy, and financial resources towards Segment A clients.

	Customer Segment A	Customer Segment B
Avg. Revenue per Transaction	$75	$100
Labor, Fulfillment, and Restocks	$12	$25
Shipping Costs	$10	$30
Customers Service Costs	$6	$12
Total Costs	**$28**	**$67**
Revenue-Costs	$47	$33

Source: - *(DAN LEBLANC, 2023)*

It would have been impossible to have a true understanding of the profitability of Segment A in comparison to Segment B without doing a customer profitability study. In the context of your actual business, there are a tone of possible insights that you may uncover to improve consumer knowledge and spend more efficiently.

6.6 Security and Privacy Concerns in Financial Cloud BI

Although cloud BI offers strong data analytic capabilities, it also poses privacy and security issues. Businesses using cloud-based business intelligence tools must manage a wide range of risks, from data breaches to regulatory issues.

The shared responsibility approach is essential to comprehending cloud BI security. Customers must safeguard their data and apps while suppliers secure the infrastructure. Protecting sensitive data in the cloud requires encryption, access restrictions, and compliance procedures (fiveable, 2023).

Data Security and Privacy in Cloud BI

Key security and privacy concerns

- **Data breaches and unauthorized access**
  - Sensitive data exposure due to misconfigured cloud services (unsecured S3 buckets)
  - Insider threats from cloud service provider employees with privileged access
- **Data loss and integrity issues**
  - Accidental deletion or modification of data by users or administrators
  - Hardware failures or natural disasters (earthquakes, floods) affecting cloud infrastructure

- **Lack of visibility and control over data**

  - Difficulty in monitoring data access and usage across multiple cloud services

  - Limited ability to enforce company-specific security policies in shared cloud environments

- **Compliance and regulatory challenges**

  - Ensuring compliance with industry-specific laws (such as GDPR for personal data and HIPAA for healthcare)

  - Maintaining data sovereignty and cross-border data transfer restrictions (EU-US Privacy Shield)

Shared responsibility model

- Assignment of security duties between the customer and the cloud service provider

  - Cloud service provider secures the underlying infrastructure and services

    - Data centres' physical security (surveillance, access restrictions)

    - Network security and access controls (firewalls, intrusion detection)

    - Patching and updating of cloud infrastructure (operating systems, virtualization)

  - Customer responsible for securing their data and applications

    - Configuring access controls and permissions (user roles, privileges)

    - Sensitive information is encrypted both in transit and at rest.

    - Tracking user behaviour and data access (audit logs, anomaly detection)

- Importance of understanding the specific responsibilities for each cloud service model
  - Infrastructure as a Service (IaaS) gives customer more control and responsibility (Amazon EC2)
  - Platform as a Service (PaaS) involves shared responsibility between provider and customer (Microsoft Azure)
  - Software as a Service (SaaS) has provider handling most security aspects, customer manages user access (Salesforce)

Encryption and access controls

- Data encryption
  - Protects data confidentiality and integrity
    - Encrypts data at rest (stored in the cloud)
    - Encrypts data in transit (during transmission)
  - Reduces risks of an organization's data being accessed or stolen by an unauthorized person.
  - Assures that the company observes legal policies (HIPAA, PCI DSS)
- Access controls
  - Implement strong authentication mechanisms
    - Multi-factor authentication (MFA) using tokens or biometrics
    - Single sign-on (SSO) for centralized access management across applications
- Apply granular access permissions
  - Permissions based on people's functional responsibilities known as, role-based access control (RBAC).
  - Policy of giving minimum access rights needed to perform an operation.

- Monitor and audit user activity
  - Detect and investigate suspicious access attempts (brute-force attacks)
  - Maintain audit logs for compliance and forensic purposes (user actions, timestamps)

Compliance and regulations

- Industry-specific regulations
  - HIPAA for healthcare data to protect the Insurance Portability and Accountability Act data
  - General Data Protection Regulation (GDPR) for the protection of people's personal data
  - GDPR (General Data Protection Regulation) for properly protecting personal data of common EU citizens
- Data sovereignty and localization requirements
  - Splitting control of data to specify where that data is stored and processed (data localization)
  - Complying with country-specific data protection laws (China's Cybersecurity Law)
- Conducting regular compliance audits
  - Assessing the effectiveness of security controls (penetration testing, vulnerability scans)
  - Identifying and addressing gaps in compliance (remediation plans, risk assessments)
- Obtaining relevant certifications and attestations
  - SOC 2 (Service Organization Control 2) for demonstrating security, availability, and confidentiality
  - ISO 17799 For IS / IT Security: Information security management based on the ISO 27001

6.7 Case Study: Bank's Adoption of Cloud BI Solutions

A prominent global bank, with operations spanning multiple regions, faced significant challenges in managing its data infrastructure. The bank's legacy systems were siloed and inefficient, making it difficult to analyze vast datasets for actionable insights. Some of these constraints made it difficult for it to effectively identify fraud in real-time, compliance to the required regulatory standards and offer its services to different customers with different needs. Understanding that the approach applied up until then was backward, the bank started working on implementing a Cloud Business Intelligence (Cloud BI) solution.

Implementation of Cloud BI

The bank teamed up with one of the top Cloud BI Solution providers to bring the departments and geographical locations data in a unified, elastic, and cost-effective cloud solution. Such a platform allowed structuring and unstructuring of data so as to avail data ingestion for analytics and reporting in real time. Primary areas that focused at the implementation included:

- **Real-Time Fraud Detection:** By integrating machine learning models into the Cloud BI platform, the bank developed predictive algorithms to identify suspicious activities. For instance, if its' transactions profile or behaviours of its customers deviate from the averaged patterns, the deviation's immediate alert was possible to combat fraud.

- **Regulatory Compliance and Reporting:** The solution automated compliance reporting, reducing manual efforts and errors. Real-time data feed and ready-made templates also guaranteed that the bank complied with various regulations in different locations and was transparent to the regulators.

- **Customer Segmentation and Profitability Analysis:** To this effect, through Cloud BI's smart analytics, the bank categorized the customers according to how they spent, historic credit records, and rate of interaction. It also enabled it to develop suitable financial products and supervise campaign that retained its customers and was lucrative.

Addressing Security and Privacy Concerns

First of all, it became obvious that trustees had some critical concerns about using databases, which were directly connected with data protection concerns, and these concerns can be explained, firstly, by the fact that financial information is more vulnerable than other types of data. Therefore, to deal with this, the Cloud BI platform put in place stringent encryption mechanisms and stringent user access mechanisms and anonymized the data. The solution also complied with international datasets like GDPR and CCPA in order to guarantee the customers' trust.

Results and Impact

The use of Cloud BI brought significant changes in the bank as mentioned next:

- **Enhanced Efficiency:** Real-time analytics reduced processing times for fraud detection and compliance reporting by 50%, significantly improving operational efficiency.

- **Improved Decision-Making:** Through the two dashboards and the models created, top managers received comprehensive insights, which allowed them to make decisions ahead of time.

- **Customer Satisfaction:** Implementation of the principles of personalized offers and faster response time grew the rates of customer satisfaction by 25%, thus improving profitability.

- **Scalability and Agility:** This characteristic helps the bank enjoy flexibility and expansion as Cloud Computing facilitates the management of a large amount of data and introduces more services into the market.

This case study is, therefore eye-opener in establishing Cloud BI as a solutions provider in the financial sector. When the bank adopted new-generation technology to improve its data handling mechanisms, it was able to respond to important operational needs also establish itself as an innovative centre in the context of the financial market.

6.8 Chapter Summary

This chapter unfolded how the Cloud BI paves the way for transformations in security compliance and customer-oriented operations. It started with the presentation of the shared responsibility model as the structure for the relationships between cloud users and cloud providers, where the former must monitor the latter while remaining accountable for their data and applications. Cloud risk management was identified as key, together with others, including security assessments, the use of AWS WAF and Microsoft Azure Firewall, and compliance with rules including PCI-DSS, ISO 27001, SOX, and GDPR.

The chapter provided key compliance requirements for the laws like CCPA, HIPAA, and FedRAMP, and the emphasis was on the strong security measures besides privacy preservation, and constant compliance checks. It pointed out that it takes more than just governance frameworks, tools and setting benchmarks and measures to attain and sustain cloud compliance all the time.

The discussion also incorporated a close look at the areas of customer segmentation and profitability analysis as key components of Cloud BI. Concepts such as behavior-based, clustering, and predictive programs help organizations target specific customer groups leading to increased commitment, satisfaction and profitability. In the context of presenting the subject of these strategies and options, the chapter laid emphasis on a competitive approach to revenues and costs with specific reference to the balances needed in customer interactively.

Last, the chapter has given a case let of a bank where cloud BI has been implemented to get rid of fraud detection, compliance and customer profitability issues. The outcome was increased operational effectiveness, improved decision-making processes, and increased flexibility of the organization.

In conclusion, the chapter showed how Cloud BI can help organisations to improve security, compliance, and customer perspectives and help organisations to cope with regulatory issues, maximize resources, and provide value-driven, data-backed decisions to reinvent the business environment in today's complex world.

Multiple Choice Questions (MCQs)

1. **What is the primary advantage of using cloud-based analytics for risk management in financial services?**

   a. Lower data storage costs

   b. Improved scalability and flexibility in risk analysis

   c. Reduced data processing speed

   d. Increased dependency on on-premise hardware

2. **Which of the following is a key benefit of fraud detection and prevention with cloud-based BI?**

   a. Increased manual intervention

   b. Real-time monitoring and analytics

   c. Limited access to data

   d. Inability to handle large data volumes

3. **How does cloud-based BI help with regulatory compliance reporting in the financial industry?**

   a. By reducing the need for manual audits

   b. By automating data collection and reporting processes

   c. By making data storage less secure

   d. By increasing data redundancy

4. **Which of the following is a key application of customer segmentation in cloud BI for financial services?**

    a. Simplifying fraud detection processes

    b. Identifying high-value customers for targeted marketing

    c. Monitoring system performance

    d. Reducing compliance reporting workload

5. **What is a primary concern when using cloud-based BI in financial services?**

    a. Enhanced data accessibility

    b. Security and privacy risks

    c. Improved decision-making capabilities

    d. Increased customer satisfaction

6. **In a cloud-based BI solution, which aspect is most critical to ensure compliance with data privacy regulations?**

    a. Data encryption and secure access controls

    b. Frequent system updates

    c. High-speed internet connection

    d. Regular customer feedback

7. **What type of analysis can cloud BI facilitate to improve financial service profitability?**

    a. Customer segmentation

    b. Cost-cutting measures

    c. Employee performance reviews

    d. Operational inefficiencies

8. **Which of the following is a typical outcome of implementing cloud-based BI for fraud detection?**

    a. Increased operational costs

    b. Faster detection and response to suspicious activities

    c. Decreased data security

    d. More manual interventions in data analysis

9. **What role does cloud-based BI play in a financial institution's regulatory compliance?**

    a. It reduces the need for external audits.

    b. It allows real-time tracking and reporting of compliance metrics.

    c. It increases the risk of non-compliance.

    d. It removes the need for regulatory training.

10. **Which of the following is an example of a security concern when adopting cloud BI in financial services?**

    a. Increased operational efficiency

    b. Exposure to cyberattacks or data breaches

    c. Enhanced customer experience

    d. Greater control over data access

Answer

1	2	3	4	5	6	7	8	9	10
b	b	b	b	b	a	a	b	b	b

BIBLIOGRAPHY

A-khateeb, B. (2024). Business Intelligence (BI): *International Journal of Asian Business and Information Management, 15*, 1–15. https://doi.org/10.4018/IJABIM.340387

aceinfoway. (2024). *How Manufacturers Can Enhance Quality Control by 25% with Cloud-Based Analytics.* Aceinfoway.

Adebunmi Okechukwu Adewusi, Ugochukwu Ikechukwu Okoli, Ejuma Adaga, Temidayo Olorunsogo, Onyeka Franca Asuzu, & Donald Obinna Daraojimba. (2024). BUSINESS INTELLIGENCE IN THE ERA OF BIG DATA: A REVIEW OF ANALYTICAL TOOLS AND COMPETITIVE ADVANTAGE. *Computer Science & IT Research Journal.* https://doi.org/10.51594/csitrj.v5i2.791

Ahmed, A. A., & Hussan, M. I. T. (2018). Cloud Computing: Study of Security Issues and Research Challenges. *International Journal of Advanced Research in Computer Engineering & Technology (IJARCET).*

Akshay Badkar. (2023). *Guide to Cloud-based Inventory Management.* Simplilearn.

Al-Aqrabi, H., Liu, L., Hill, R., & Antonopoulos, N. (2015). Cloud BI: Future of business intelligence in the Cloud. *Journal of Computer and System Sciences.* https://doi.org/10.1016/j.jcss.2014.06.013

Al-marsy, A., Chaudhary, P., & Rodger, J. A. (2021). A model for examining challenges and opportunities in use of cloud computing for health information systems. *Applied System Innovation.* https://doi.org/10.3390/asi4010015

aquasec. (2023). *8 Cloud Compliance Standards & 7 Steps to Achieving Compliance*. Aquasec.

Aron powers. (2023). *Building Scalable Data Architectures: Principles and Best Practices*.

Atlan. (2023). *Data Quality in Data Governance: The Crucial Link That Ensures Data Accuracy and Integrity*.

Automation, S. (2022). *A Complete Guide to Regulatory Compliance in Healthcare*. Scrut Automation. https://www.scrut.io/post/regulatory-compliance-in-healthcare

Balachandran, B. M., & Prasad, S. (2017). Challenges and Benefits of Deploying Big Data Analytics in the Cloud for Business Intelligence. *Procedia Computer Science*. https://doi.org/10.1016/j.procs.2017.08.138

Bryan Christiansen. (2023). *3 Smart Manufacturing Use Cases That Improve Production*. Rtinsights.

Chatterjee, D. (2024). *Episode 63 -- Securing Application Programming Interfaces (APIs)*.

Chen, H., Chiang, R. H. L., & Storey, V. C. (2012). Business intelligence and analytics: From big data to big impact. *MIS Quarterly: Management Information Systems*. https://doi.org/10.2307/41703503

cloud. (2024). *What is cloud data security? Benefits and solutions*. Cloud.

Coats, B., & Acharya, S. (2014). Leveraging the cloud for electronic health record access. *Perspectives in Health Information Management / AHIMA, American Health Information Management Association*.

DAN LEBLANC. (2023). *What is Customer Profitability Analysis? | Complete Guide*. Daasity.

Dash, S., Shakyawar, S. K., Sharma, M., & Kaushik, S. (2019). Big data in healthcare: management, analysis and future prospects. *Journal of Big Data*. https://doi.org/10.1186/s40537-019-0217-0

Davidiseminger, Mohitp, Court, TimShererWithAquent, V-kents, Mihart, & V-hearya. (2023). *Use the Analytics pane in Power BI Desktop.* Microsoft. https://learn.microsoft.com/en-us/power-bi/transform-model/desktop-analytics-pane

Disney, W. (2023). *5 Methods for Creativity.*

DOMO. (2023). *What are data connectors and why are they essential for businesses?* DOMO. https://www.domo.com/learn/article/what-are-data-connectors-and-why-are-they-essential-for-businesses

Education, J. (2024). *5 Powerful Benefits Of Inventory Management.* Jaro Education. https://www.jaroeducation.com/blog/benefits-of-inventory-management/

FedRAMP. (2020). *FedRAMP.* FedRAMP. https://www.gsa.gov/technology/government-it-initiatives/fedramp

fintrak. (2023). *Security and Privacy Considerations for Cloud BI.*

fiveable. (2023). *Data Security and Privacy in Cloud BI.* Fiveable.

Gangwar, H. (2017). Cloud computing usage and its effect on organizational performance. *Human Systems Management.* https://doi.org/10.3233/HSM-171625

GENPACT. (2024). *Four steps to improving the customer experience with data and analytics.* GENPACT.

Grow.com. (2024). *Data Quality Counts: How To Ensure Accuracy in Your BI Reporting Tools.*

Hyperproof. (2022). *The Sarbanes-Oxley Act (SOX).* Hyperproof. https://hyperproof.io/sarbanes-oxley-act/

impact. (2024). *Data Analytics in Manufacturing.* Impact.

intwo. (2024). *Cloud transformation challenges in manufacturing.* Intwo.

Isms. (2022). *Securing Card Transactions Using Payment Card Industry Data Security Standard (PCI DSS).* Isms Online. https://www.isms.online/pci-dss/

johns hopkings. (2024). *The National Institute of Standards and Technology (NIST)*. Johns Hopkings. https://engineering.jhu.edu/mtaheri/MURI/research/national-institute-of-standards-and-technology/

Jonny Parker. (2024). *Cloud-based inventory management 101: A quick introduction*. Fishbowlinventory.

Karkouch, A., Mousannif, H., Al Moatassime, H., & Noel, T. (2016). Data quality in internet of things: A state-of-the-art survey. *Journal of Network and Computer Applications, 73*, 57–81. https://doi.org/10.1016/j.jnca.2016.08.002

Kasem, M., & E. Hassanein, E. (2014). Cloud Business Intelligence Survey. *International Journal of Computer Applications*. https://doi.org/10.5120/15540-4266

Khan, S., Khan, H. U., & Nazir, S. (2022). Systematic analysis of healthcare big data analytics for efficient care and disease diagnosing. *Scientific Reports*. https://doi.org/10.1038/s41598-022-26090-5

korcomptenz. (2024). *Cloud-Based Fraud Prevention Solutions and Prevention in Banking - Enhancing Security and Trust*. Korcomptenz.

Kujawski, M. (n.d.). *Building an Efficient ETL/ELT Process for Data Delivery*. 2024.

Lee, C. S., Cheang, P. Y. S., & Moslehpour, M. (2022). Predictive Analytics in Business Analytics: Decision Tree. *Advances in Decision Sciences*. https://doi.org/10.47654/V26Y2022I1P1-30

Loginradius. (2024). *GDPR Compliant*. Loginradius. https://www.loginradius.com/compliance-list/gdpr-compliant/

MaggiesMSFT. (2024). *Use Copilot to write and explain DAX queries*. Microsoft. https://learn.microsoft.com/en-us/power-bi/transform-model/dax-query-copilot-create

Mark R. (2023). *Choosing the Right Cloud Service Model: IaaS, PaaS, or SaaS*.

Matt Pacheco. (2024). *Cloud Risk Management: How to Identify and Mitigate Risks*. Tierpoint.

medium. (2024). *Integrating IoT Data Streams into Business Analytics Software for Enhanced Insights*. Medium.

Murray-Watson, R. (2023). *2022 Healthcare Data Breach Report*. THE HIPAA JOURNAL. https://www.hipaajournal.com/2022-healthcare-data-breach-report/

Ngu, A. H., Gutierrez, M., Metsis, V., Nepal, S., & Sheng, Q. Z. (2017). IoT Middleware: A Survey on Issues and Enabling Technologies. *IEEE Internet of Things Journal*. https://doi.org/10.1109/JIOT.2016.2615180

Niu, Y., Ying, L., Yang, J., Bao, M., & Sivaparthipan, C. B. (2021). Organizational business intelligence and decision making using big data analytics. *Information Processing and Management*. https://doi.org/10.1016/j.ipm.2021.102725

Norgeot, B., Muenzen, K., Peterson, T. A., Fan, X., Glicksberg, B. S., Schenk, G., Rutenberg, E., Oskotsky, B., Sirota, M., Yazdany, J., Schmajuk, G., Ludwig, D., Goldstein, T., & Butte, A. J. (2020). Protected Health Information filter (Philter): accurately and securely de-identifying free-text clinical notes. *Npj Digital Medicine*. https://doi.org/10.1038/s41746-020-0258-y

Olexová, C. (2014). Business intelligence adoption: A case study in the retail chain. *WSEAS Transactions on Business and Economics*, *11*(1), 95–106.

Optimove. (2024). *Customer Segmentation*. Optimove.

Paci, A. (2024). *How IT Can Enhance Customer Experience Through Data Analytics*. Linkedin. https://www.linkedin.com/pulse/how-can-enhance-customer-experience-through-data-analytics-alain-paci-bnose

Patadiya, J. (2024). *Data Security in Cloud Computing: Understand its Various Aspects and Protecting Your Data*. Radix. https://radixweb.com/blog/database-security-cloud-environment

Petrova, B. (2024). *Predictive Analytics in Healthcare*. Reveal. https://www.revealbi.io/blog/predictive-analytics-in-healthcare

Potter, K., & Broklyn, P. (2024). AI-BASED PREDICTIVE MAINTENANCE IN MANUFACTURING INDUSTRIES Background: AI-Based Predictive Maintenance in Manufacturing Industries. *Artificial Intelligence*.

Praful Bharadiya, J. (2023). A Comparative Study of Business Intelligence and Artificial Intelligence with Big Data Analytics. *American Journal of Artificial Intelligence*. https://doi.org/10.11648/j.ajai.20230701.14

radware. (2022). *Cloud Security: Principles, Solutions, and Architectures*.

Rene Abraham, Jan vom Brocke, J. S. (2019). *Data Governance: A conceptual framework, structured review, and research agenda*.

Sandhu, A. K. (2022). Big Data with Cloud Computing: Discussions and Challenges. *Big Data Mining and Analytics*. https://doi.org/10.26599/BDMA.2021.9020016

Sansico, T. (2019). *How Can Business Intelligence Leaders Adopt CCPA Compliance?* Wisdom. https://wiiisdom.com/blog/business-intelligence-ccpa-compliance/

sensemore. (2024). *7 Essential Strategies for Implementing Predictive Maintenance*. Sensemore.

SG Analytics. (2024). *The Role of AI and Data Analytics to Drive Personalization Strategies*. SG Analytics.

Shah, K. (2017). *Performance Optimization Techniques*. Thirdrocktechkno. https://www.thirdrocktechkno.com/blog/performance-optimization-techniques/

Sharma, N. (2024). *How Business Intelligence for Supply Chain Management Helps Businesses Optimize Operations and Drive Growth*. Appinventiv.

shopify. (2024). *What Is Cloud-Based Inventory Management? A Complete Guide*. Shopify.

Signatech. (2022). *What Are the Benefits of Cloud Business Intelligence Solutions?* Signatech. https://signatech.com/blog/benefits-of-cloud-based-business-intelligence-solutions/

Signatech. (2023). *How to Overcome Common Challenges in Cloud BI Implementation?* Signatech. https://signatech.com/blog/cloud-bi-implementation-tips/

Sinaga-Bulgamin, S. (2022). *Research of Point-of-Sale Systems for Integrating International Business Operations.*

Solutions, G. (2023). *Enhancing Digital Customer Experiences Through Data-Driven Insights*. SIDGS.

Sun, Z., Sun, L., & Strang, K. (2018). Big Data Analytics Services for Enhancing Business Intelligence. *Journal of Computer Information Systems*. https://doi.org/10.1080/08874417.2016.1220239

Tableau. (2020). *Business intelligence: A complete overview*. Tableau. https://www.tableau.com/business-intelligence/what-is-business-intelligence#:~:text=Further learning-,What is business intelligence%3F,make more data-driven decisions.

Team, Y. (2022). *The History and Evolution of Business Intelligence (BI) Platforms*. Yellowfin Team. https://www.yellowfinbi.com/blog/the-history-and-evolution-of-business-intelligence-platforms

Tech target. (2024). *What is data architecture? A data management blueprint.*

Tiwari, D. (2024). Role of Data Analytics in Business Decision Making. *Knowledgeable Research: A Multidisciplinary Journal, 3*, 18–27. https://doi.org/10.57067/0zr57x43

TOADER, A.-G. (2022). Cloud-Computing. *RESEARCH FOCUS, 47*.

Wang, Y., Kung, L. A., Wang, W. Y. C., & Cegielski, C. G. (2018). An integrated big data analytics-enabled transformation model:

Application to health care. *Information and Management.* https://doi. org/10.1016/j.im.2017.04.001

Ware, C. (2024). *HIPAA Compliance: What is it & How do I become Compliant?* Currentware. https://www.currentware.com/blog/hipaa-compliance-what-is-it-how-do-i-become-compliant/

World Health Organization, OECD, & World Bank. (2018). Delivering quality health services: a global imperative for universal health coverage. Geneva: ; 2018. In *World Health Organization, World Bank Group, OECD.*

Zeotap. (2024). *Zeotap Awarded ISO 27001 Certification for Information Security Management.* Zeotap. https://zeotap.com/news/zeotap-awarded-iso-27001-certification-for-information-security-management/

ABOUT THE AUTHOR

Sethu Sesha Synam Neeli is a highly experienced database architect and engineer with over 15 years of work experience in changing the ways industries address data management and data security along with effective data usage. Neeli was actively involved in development of database systems for manufacturing and e-commerce sectors, cloud solutions and disaster recovery as well as data warehousing solutions for public service sector.

Some of the technical skills that come out clearly in the work Neeli has done include designing large or large-scale database systems. His expertise in improving the data pipeline and leveraging AWS Redshift, Aurora, and Azure SQL Database for advancing the companies have benefited McDonald's, Toyota Motors North America and Microsoft.

Being a performance tuning specialist in databases, Neeli has always had a record of performance, for instance, she had an achievement of increasing the query response rate by 25% at Market America. The strong HA/DR frameworks and security, including encryption and auditing as well as secure coding, provide the necessary levels of security of important platforms.

Besides industry contributions, Neeli is a distinguished researcher and an opinion maker with research papers on edge cloud computing, federated learning, and AI-based query optimization. He uses it in open-source development, too, having built tools such as DBSecureManager and DBMigrationEase in an attempt to improve database security, migration, and performance.

Neeli's legacy is of innovation, perfecting the art of technical skills and at the same time being a team player and someone who researches. It keeps on extending benchmarks in database technologies so as to help various businesses and communities flourished in the present world.

www.ingramcontent.com/pod-product-compliance
Lightning Source LLC
Chambersburg PA
CBHW040726120726

48010CB00001B/22